· THE ·
LIVING WORLD

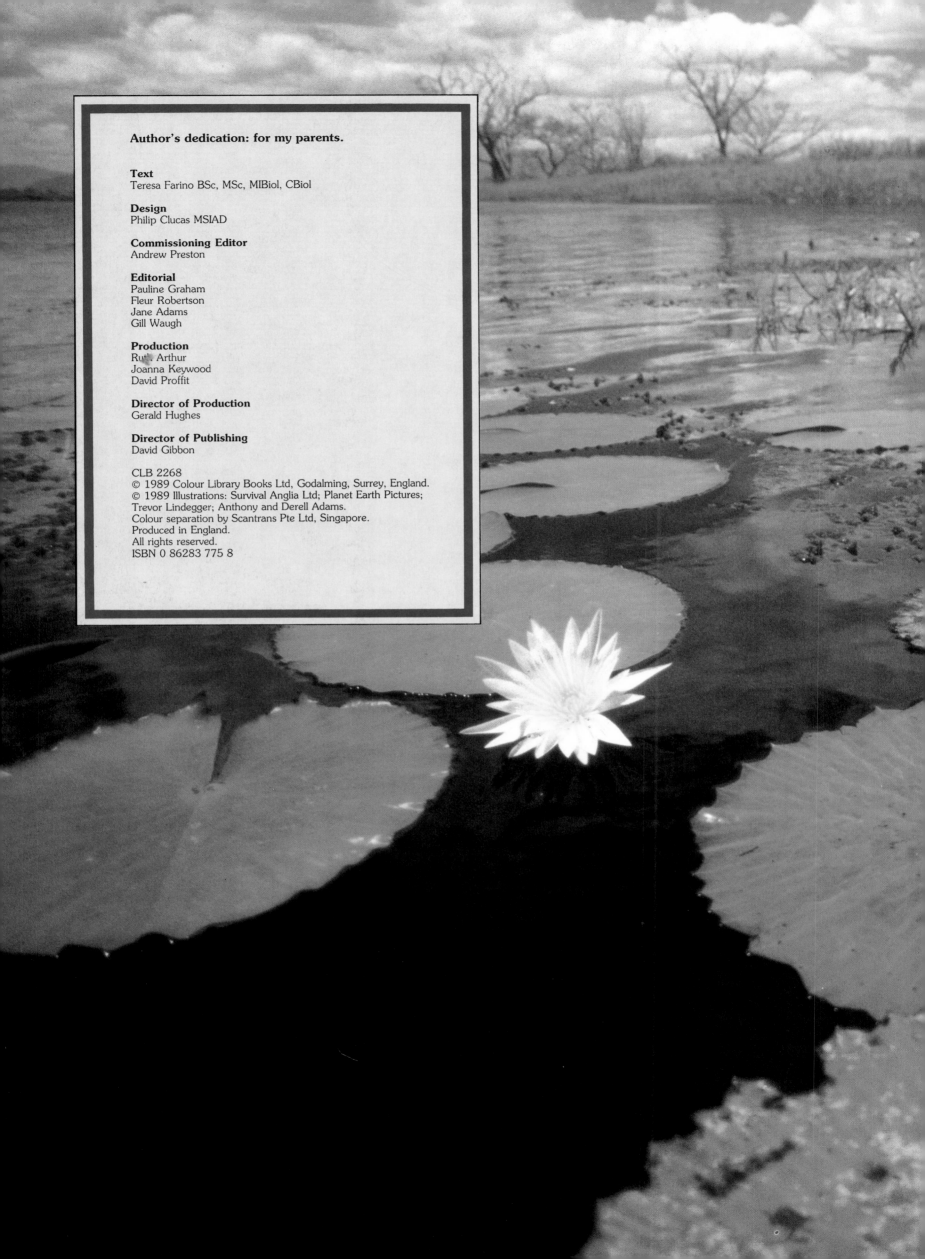

Author's dedication: for my parents.

Text
Teresa Farino BSc, MSc, MIBiol, CBiol

Design
Philip Clucas MSIAD

Commissioning Editor
Andrew Preston

Editorial
Pauline Graham
Fleur Robertson
Jane Adams
Gill Waugh

Production
Ruth Arthur
Joanna Keywood
David Proffit

Director of Production
Gerald Hughes

Director of Publishing
David Gibbon

CLB 2268
© 1989 Colour Library Books Ltd, Godalming, Surrey, England.
© 1989 Illustrations: Survival Anglia Ltd; Planet Earth Pictures;
Trevor Lindegger; Anthony and Derell Adams.
Colour separation by Scantrans Pte Ltd, Singapore.
Produced in England.
All rights reserved.
ISBN 0 86283 775 8

THE LIVING WORLD

Teresa Farino

Bramley Books

· CONTENTS ·

Previous pages: waterlilies on Lake Malawi, (inset left) a bee orchid and
(inset right) a red-eyed tree frog. Left: the translucent magenta polyps of
the sea anemone *Corynactis* and (above) an Indonesian Atlas moth.

INTRODUCTION

Our Earth is the only planet in the solar system – possibly in the universe – where conditions exist in which life as we know it can thrive. Viewed from space it is one of the most beautiful sights ever beheld by man: a blue and green jewel wreathed in tendrils of cloud, shimmering in the vast blackness of eternity.

When the world was born, around 4,500 million years ago, it must have been a bleak and desolate land of cooling rock and violent volcanic activity. As the contortions diminished, water began to accumulate in depressions in the Earth's crust and the oceans were formed. The odds against the chemical accident from which all life is descended are phenomenal, and yet in the primeval oceans the first simple organisms – probably rather like modern-day bacteria – sprang into existence.

Without the sun there would be no life on this planet, but its gift of warmth and light is a double-edged sword, since it also gives out strong ultraviolet rays which are inimical to animal life. For millennia the only life-forms which were able to exist outside the aquatic environment were plants. As each plant harnessed energy from the sun in order to grow, and released oxygen during the process of photosynthesis, a layer of ozone formed in the atmosphere which prevented much of the harmful ultraviolet radiation from reaching the Earth. Under the protection of this invisible mantle, animal life began to ascend the ancient shores of the oceans and lakes, gradually evolving into true land creatures.

From these humble beginnings animals and plants have come to occupy the four corners of the Earth. Even in the frozen wastes of the polar regions and the scorching sands of the great deserts there is life – bizarre animals and plants adapted to the vagaries of their environments. Although countless species became extinct during the history of the planet, living organisms continued to meet every challenge that the world could throw at them: drifting continents, volcanoes, mountain upheaval, the birth and death of oceans, climatic changes. With each change the balance shifted a little, but equilibrium was eventually restored. Until now. Now one species has evolved that threatens to destroy the fragile natural balance of the planet – man.

Facing page: the splendid umbellifer, Queen of the Alps, (top) South America's common iguana, (above left) African lions and (above right) the magnificent bald eagle.

Even lion cubs (**10**) are vulnerable to predators as the mother must leave them to drink and hunt. Their camouflaging spotted coats provide some protection on these occasions. A skilled hunter of reptiles, particularly snakes, the short-toed eagle (**4**) breeds in southern Europe and north Africa, moving further south for the winter, as does the insectivorous whinchat (**5**). Bearded chameleons (**6**), slow-moving arboreal lizards, trap invertebrates with their long, sticky, extensible tongues, themselves avoiding predators by changing colour according to their setting. The bald head and neck of the marabou (**3**), the largest of the world's storks, are ideally suited to its scavenging lifestyle.

7

8

9

eventually lead to arid, desert-like conditions. In conclusion, the existence of the world's grasslands is dependent on several factors: climatic influences, frequent fires, grazing animals and lastly, man's recent interference with the earth's natural vegetation.

The principal tracts of grassland in the world today are the prairies, which cover the Great Plains of America and stretch from Texas to mid-Canada; the Eurasian steppes, which range from Hungary to Mongolia and include large parts of the Soviet Union; the tree-studded savannas of Africa; the Australian grasslands, which surround the arid interior of this island continent, and the South American grasslands, 10 which comprise the pampas of Uruguay and Argentina and the llanos of Venezuela and central and southern Brazil. All grasslands can be divided into two types, temperate or tropical, according to climate.

The world's tropical grasslands, comprising mainly the African savannas and the Venezuelan llanos, lie in a belt on either side of the equator, between the tropics of Capricorn and Cancer, where there are usually three to seven rain-free months per year. The temperate grasslands occur to the north and south of this zone and, because they are further from the equator, their climate is harsher and there is a great range in both

rain falls between September and May, but there is almost none throughout the rest of the year, when daytime temperatures can exceed 35°C. It is perhaps not surprising, therefore, to discover that this is a landscape shaped largely by fire.

Even further south, in the temperate zone of South America, lie the pampas of Argentina and Uruguay. Although a few areas were planted with cereals by early white settlers, the pampas still look today much as they must have done to the explorers of the New World several hundred years ago. They stretch for over 500,000 square kilometres in a huge, unbroken sea of

The spectacular lily *Grinum zeylanicum* (**2**), a typical grassland species of tropical Asia and Africa, may reach one-and-a-half metres in height, taking advantage of the spring rains to produce up to twenty red-and-white-striped flowers.

1

daily and annual temperatures. These temperate grasslands include the prairies and steppes in the northern hemisphere and the Argentinian pampas in the southern hemisphere.

The South American grasslands all lie between the great mountain ridge of the Andes and the Atlantic Ocean. In the north of the region, the grasslands lie within the tropical zone and include the llanos of the Orinoco Basin of Venezuela and Colombia. These consist of alluvial plains where the silts and gravels provide so little in the way of root support that they are often treeless. Seasonal rains, commencing in April, often flood large areas in the low-lying depressions. These llanos are thus flooded during the summer but parched during the winter, and several animals have evolved specifically to cope with these conditions.

South of the llanos, and on the other side of the equator, lie the Brazilian cerrados. These essentially dry grasslands, with a high density of trees, cover an area in excess of one and a half million square kilometres – fifteen to twenty-five per cent of the country. There are three different types of cerrado characterised by varying amounts of tree cover, of which the campos are the most open. In addition, gallery forests line the watercourses and valleys. About 1,500 millimetres of

2

The primary function of an antelope's horns is that of defence, 200-kilogramme oryx (**1**) being quite capable of tossing or impaling large predators, but they are also important during ritual fighting to determine dominance among the males. The grass snake (**3**) lives in damp conditions right across Europe as far east as Lake Baikal and southwards to northwest Africa and feeds mainly on tailless amphibians.

feathery *Stipa* grasses from the Gran Chaco in the north to the Patagonian Desert in the south.

The vegetation of the Chaco region seems to fall midway between those of the tropical rainforests of the Amazon and the true pampas grasslands. Like the llanos, it is essentially a lowland alluvial plain, parched for part of the year, but flooded during the rainy season and forming a patchwork of swamps known as esteros. Often referred to as 'desert', the undulating Patagonian Plain, although cold and dry, is nevertheless far from lifeless, being covered with typical arid land vegetation comprising grasses, dwarf shrubs and scrub. Many of

3

4

African weaverbirds are close relatives of European sparrows, albeit with rather more colourful plumage. Their name derives from their nest-building technique, which involves the interlacing of grasses and plant fibres to a specific blueprint, such as the bulbous cone of the masked weaver (**4**). Their nests are slung from acacia branches and have long entrance tunnels to prevent egg-eating snakes from reaching the contents. The red-billed quelea (**5**) is the most common weaverbird, its breeding colonies often numbering millions.

5

6

7

Although the Asian elephant (**6**) is essentially a creature of the tropical forest, the loss of much of this habitat in its Assam stronghold means that today these huge pachyderms are often seen on the open plains. The patas (**7**), or red monkey, is an inhabitant of the tropical savannas which lie between the Sahara and the West African jungle. They live in troops of up to thirty individuals led by a single dominant male, and although they are highly territorial they rarely spend two consecutive nights in the same place, covering large areas in search of food, mostly leaves and shoots, though fruit, seeds and insects are also eaten.

1

the South American mammals typical of the open plains also make their homes here. Indeed, Patagonia is thought to have been the birthplace of the giant sloths and glyptodonts that were the predecessors of many of South America's more bizarre, present-day creatures, such as anteaters and armadillos.

Unlike the various grasslands of South America, large areas of which remain relatively unscathed by man's activities, the temperate North American prairies have all but disappeared in their natural state. Before the arrival of Europeans, the prairie was the most extensive region of natural vegetation in North America, home to only a handful of native Indian tribes, who lived in equilibrium with these grasslands and their animal occupants. Now it is the 'bread-basket' of the nation a great cereal belt stretching for mile upon golden mile.

There are essentially three different types of prairie. The well-watered, tall-grass prairie to the east, stretching from Ohio to eastern Oklahoma, is dominated by tussock-forming grasses such as big bluestem, Indian

grass and prairie cord-grass. These grasses once grew high enough to conceal a man on horseback, but today the tall-grass prairies have been hard-hit by their conversion to corn belt. On the opposite side of America, on the high, semi-arid plains which lie in the rain shadow of the Rockies, grows the short-grass prairie, dominated by sward-forming grasses, such as blue-grama and buffalo grass, which usually grow to less than half a metre in height. Within the central plain lies a zone of mixed-grass prairie, with intermediate characteristics, where june grass, little bluestem, side oats grama and western wheatgrass are common.

It is estimated that, given the extensive root systems of these grasses, as much as eighty per cent of the plant material in a prairie lies beneath the soil. The roots of big bluestem, for example, often penetrate to a depth of two metres and are capable of growing more than a centimetre a day. The shoots and leaves above ground die back during the winter, leaving the growing points well below ground level, thus enabling the plant to survive surface temperatures of -40°C. Similarly,

The lean, narrow-chested cheetah (1) is probably the fastest creature on earth, being capable of reaching speeds of up to 110 kilometres per hour over short distances in pursuit of its prey. If hunting alone, the cheetah favours gazelles, particularly Thompson's, though a group may tackle even wildebeest and zebras.

3

Although the feeding habits of the Nubian vulture (**3**) and its relatives might be considered distinctly unpleasant, these bald-headed birds perform a vital role in the grassland food chain, as well as ridding the plains of rotting animal carcasses.

4

5

Weighing some six tonnes and standing almost three-and-a-half metres tall, the African elephant (**4**) is the largest living land mammal. If threatened, it may opt to charge, unfurling its ears to their full extent to appear larger and more intimidating and reaching speeds of up to forty kilometres per hour. The African ground squirrel (**5**), although protected to a certain extent by its cryptic coloration and burrowing lifestyle, is a favourite food item with many grassland raptors, snakes and small carnivores such as the mongoose. In the dry Argentinian pampas small, omnivorous canids such as the Patagonian fox (**2**) thrive on a varied diet of small vertebrates, insects, fruit and seeds.

during the frequent prairie fires, such as those employed for more than 15,000 years by the indigenous Amerinds as a means of flushing game, the temperature a metre above ground level may be an incredible 230°C, yet only a few centimetres below the surface it is near normal.

Although grasses make up the bulk of the sward, other herbaceous species are common. During the main flowering period, from April to October, the prairies are alive with colour. In order to avoid competing for the available resources, especially light, with the dominant grasses, the insect-pollinated plants have adopted various strategies. Some mature early, before the grasses have lifted their tall flowering spikes into the wind; others adopt a trailing habit, using the grasses as a trellis, and yet others have low-growing leaves, but produce an enormously long flowering stalk to attract the attention of the insects.

Prairie grasslands once covered over three and a half million square kilometres, about fifteen per cent, of North America. In the 1800s, the conversion to

2

The snake's-head fritillary (**3**) grows in riverside meadows in northern and western Europe where traditional land management consists of flooding in winter and cutting for hay in the summer. The tongue orchid (**2**) is also a plant of anthropogenic grasslands and is mainly restricted to southern Europe.

crop lands began with the breaking up of the tough sward formed by the grasses using steel ploughs, which greatly disrupted the stability of the soil. Consequently, the serious drought of the 1930s reduced huge areas of the plains to little more than drifting clouds of topsoil, unable to support a single living organism. This, the creation of the so-called Dust Bowl, was only the tip of the iceberg: a further estimated 115 million hectares of farmland have since been damaged in a

similar manner. Furthermore, soil erosion is still critical in more than one third of farmland in the USA, yet marginal land is still being appropriated for growing crops.

The other major natural grassland region of the northern hemisphere is the Eurasian steppes. Compared with the prairies, the soils here are less fertile and the vegetation is both less diverse and less abundant. This is due partly to the relentless, dry winds

The black-naped hare (**4**) of the Indian subcontinent is a characteristic herbivore of grasslands that are seasonally flooded.

26

that sweep across these plains, but mainly to the continental climate and consequent low rainfall. As a result, trees are rare. Like the American prairies, however, the natural steppe vegetation has been replaced to a large extent by wheatfields, particularly between the Polish border and the Caspian Sea.

The steppes stretch from eastern Europe to the Orient, bounded to the north by the seemingly endless coniferous forests of the taiga and to the south by the deserts of Central Asia. The flowers of the steppes – tulips, peonies, anemones, irises and buttercups – all bloom in spring; later in the year the plumed seed heads of the feather grass dominate the landscape, towering over the other grasses, such as crested hair-grass and bulbous poa. The more saline regions of the

steppes are formed where evaporation exceeds rainfall, thus drawing water and mineral salts upwards through the soil to the surface. Here halophytic, or salt-tolerant, plants, such as black and white wormwoods, grow.

The other region of natural temperate grassland lies in the southern hemisphere, in New Zealand. In 1840, grasslands covered almost half of South Island, but today only fifteen per cent of the country's total land area is natural grassland. Once again this is due largely to the agricultural activities of European settlers. The Canterbury Plains on South Island, for example, once extended for some 65,000 square kilometres. As in the prairies, the tussock-forming grasses were predominant; in fact, there are no native sward-

In the absence of chemical fertilisers and pesticides, a wealth of arable weeds, especially poppies (**1**), thrives in the cereal fields of southern Europe. Natural grasslands in North America are a riot of colour in spring, characteristic species including the Texas primrose (**9**). As a result of 5,000 years of traditional management, the haymeadows of the Picos de Europa (**8**) in northern Spain are some of the most diverse Atlantic grasslands in the world. Old meadows provide valuable nest-sites for typical birds of open habitats, such as the skylark (**6**), while their damp, shady margins provide a niche for fungi, including the lawyer's wig, or shaggy ink-cap (**7**).

6

7

8

9

forming grasses in New Zealand at all. Meadow grasses and fescues were the main species, sheltering a rich herbaceous flora between the tussocks, with the occasional cabbage tree dotting the plains. Today, this area has been completely cleared and its vegetation replaced with more productive species to provide arable crops and man-made pastures.

An estimated fifteen million square kilometres of the earth's surface are today covered in tropical savanna

The silver-spotted skipper (**5**) has an enormous range, inhabiting calcareous grasslands from Europe through temperate Asia to western North America.

grasslands, all lying within two broad tracts of land flanking the equator, where daytime temperatures rarely drop below 24°C. Tropical savanna is the term used to describe any extensive, flat grassland area in the tropical regions of the world and includes the llanos, cerrados and campos of South America, as well as those grasslands occurring on the Deccan plateau of India and in Australia's tropical belt, particularly where trees are a predominant feature of the landscape. Other parts of the world where tropical savannas occur include western Madagascar and the Mexican plateau; they even form a mosaic within the tropical rainforests of Borneo and New Guinea in Indonesia. But perhaps the best known of the savanna grasslands are those in Africa, which form the largest continuous area of this type of vegetation and cover about one third of that continent.

The African savannas lie in a belt some 4,800 kilometres long, stretching from the Atlantic to the Indian Ocean. Within this zone, two main regions of savanna can be distinguished. Lying to the north of the equator and bordering on the Sahara Desert is a broad belt known as the Sudan savanna, which stretches from the Horn of Africa westwards to southern Senegal and also includes much of Uganda and Kenya. South of the equatorial jungles a similar zone is contiguous with the arid Kalahari region, extending over much of Angola, Zambia, Tanzania, Zimbabwe, Transvaal and Mozambique; this area is referred to as veld. The most natural African savannas in existence today, that is to say, those which have suffered least disturbance by man, lie within the Serengeti National Park in Tanzania, the Nairobi National Park in Kenya and the Kabalega Falls National Park in Uganda.

As on the North American prairies, over much of the continent of Africa the grasses are perennials of the tussock-forming type, which are well-adapted to the climatic cycle and the great variation in daytime temperatures. The savanna sward is a mixture of palatable grasses and herbs. Red oat grass is often dominant, most notably on well-drained soils. The taller species, such as elephant grass, grow mainly in the depressions, into which water drains from the surrounding slopes, thus providing a more reliable year-round water supply. At the height of the growing season stands of these grasses are commonly three to four metres tall and form a dense jungle of stems; in places they may even exceed a phenomenal five

1

2

3

4

There is some evidence to suggest that orchids of the genus *Ophrys* have a bizarre flower structure that resembles a female hymenopteran in order to deceive male bees and wasps into copulating with the mimics, cross-fertilising the flower in the process. The commoner European *Ophrys* species include bee orchids (**1**), early spider orchids (**3**), sawfly orchids (**5**) and sombre bee orchids (**7**). The day-flying broad-bordered bee hawk (**2**) closely resembles a bumblebee when at rest, although the reason for this mimicry is uncertain.

The ancient tradition of haymaking in many parts of Europe favours orchids as most are early-flowering species that have set seed before cutting and have tuberous roots in which to store food reserves for the coming spring. Evocatively-named orchids, many of which favour calcareous soils, include pyramidal orchids (**6**), early purple orchids (**8**), greater butterfly orchids (**9**), lizard orchids (**10**), pink butterfly orchids (**11**) and fragrant orchids (**12**). (**4**) A typical haymeadow in northern Spain supports a dense population of heath spotted and fragrant orchids.

5

6

7

8

9

10

11

12

metres in height. In contrast, the higher, drier land is covered with the tropical equivalent of short-grass prairie, whilst the slopes shelter grasslands of mixed characteristics. The diversity of grasses is staggering: Tanzania alone boasts fifty-five different species of guinea grass and forty-three species of dropseed. Almost all savanna grasses are palatable, the more so when they are young and tender, as is the case when grazing pressure is sufficient to encourage constant new leaf growth.

The grasslands with which the majority of Europeans are most familiar are those which were carved from the forests of Europe by their ancestors and have since been maintained as such by grazing, mowing or fire. About 7,000 years ago much of Europe was covered with forests of oak, mixed with elms and limes. Yet only 2,000 years later, both Neolithic and early Bronze-Age peoples were clearing these woodlands to provide pasture for their domestic herds and were beginning to practise the sort of 'slash and burn' crop-growing techniques that the Amazon Indians still employed

1

2

3

4

5

until recently. The chalk grasslands of southern England and northern France, for example, were among the first to emerge from the virgin forests of Europe. They have been maintained not by wild herbivores, but through their use by man as pasture for his sheep; fire has also played its part.

Without doubt these rolling chalk pastures support very varied communities of plants and invertebrates, particularly butterflies. Nevertheless, the bird and mammal life of these open downlands consist almost entirely of species with a wide ecological tolerance, that is, those that can live in many different habitats.

6

Pasqueflowers (**3**) and masterwort (**8**) both thrive on lime-rich soils, and old meadows are often dominated by sheets of greater yellow rattle (**1**), which is semi-parasitic on grasses. The more diverse a grassland botanically, the richer its butterfly fauna; the larvae of the scarce copper (**7**) feed on docks, and those of the painted lady (**6**) on thistles and nettles. The life-cycle of the large blue (**5**) necessitates a symbiotic relationship with the meadow ant and the presence of wild thyme.

This may be attributed to the relatively short period of time these grasslands have been in existence, compared with the African savannas and North American prairies. Thus the larger, slower-breeding animals have not yet had sufficient time to develop specialisations that would enable them to exploit this new habitat to its full. Indeed, because of the rapidly increasing human population, it is unlikely that such evolutionary changes will ever be allowed to occur. Chalk grasslands, so recently after their creation by the ancestors of modern man, are suffering the same fate as the primeval prairie and steppe grasslands elsewhere in the northern hemisphere. They are being cultivated to feed the growing mass of humanity.

There are essentially two types of man-made

7

8

9

10

The kestrel (2) is a small falcon whose breeding range extends from Europe across Asia and into northern Africa. When hunting, it hovers almost motionless, scanning the grassland for insects and small vertebrates, including the newborn chicks of the lapwing, or peewit (4), which favours marshy meadows with short turf as a breeding ground. The white-footed mouse (10) is a gregarious rodent of the North American prairies, and the metallic green little bee-eater (9) of Africa is possibly the most numerous of all twenty-four members of its family.

grassland. Pastures are utilised in the same way as the natural grasslands of the world, to produce protein in the form of meat, in this case using domesticated herbivores such as sheep, cattle and goats. Meadows, on the other hand, are grasslands from which man takes a summer hay crop with which to feed his animals during the winter, when the grasses are dormant and do not grow. Ideally, he does this just before the grasses seed, when their nutritional content is highest.

The herbs which grow in pastures and meadows have adopted different survival strategies. Pasture plants are low-growing, often in the form of rosettes,

to avoid destruction of the buds by grazing animals; they may also possess spines or hairs to deter herbivores. Meadow species, on the other hand, usually grow to a much greater height, having nothing to fear from grazing animals; however, to ensure their survival, they flower and set seed early, before the haymaking season. These haymeadows occur in many parts of Europe, but nowhere are they quite so rich and colourful as in northern Spain, where man still employs traditional methods of animal husbandry.

It has been estimated that several centuries ago over forty per cent of the earth's land surface was covered with natural grasslands, but today only about half of these remain. Man has exploited the vast prairies and steppes for his own needs, growing cereals and other crops and replacing the native grasses with more productive strains, the better to feed his rapidly multiplying domestic herds. In Eurasia only about sa hundred square kilometres of virgin steppe grassland remain. In North America, two thirds of Canada's prairie lands have been lost, together with about half of the original United States' prairies. Illinois, once known as the Prairie State, had 145,000 square kilometres of prairie 200 years ago; this has been so decimated over the years that only a few hectares remain untouched today. In addition, most of the large herbivores that have been closely linked with these natural grasslands for millions of years have been reduced to a few meagre populations in remote corners of the world.

Bison are essentially wild cattle, although an adult bull may stand two metres high at the shoulder and

1

The variety of feeding habits adopted by savanna herbivores alleviates competition between species. Both impalas (**1**) and gerenuks (**2**) are essentially browsers, the latter adept at standing on its hind legs to nibble at acacia branches, while springboks (**3**) tend towards a grazing lifestyle and wildebeest (**4**) take the succulent stems of plains grasses after zebras have eaten the coarser tops.

2

3

4

weigh 1,000 kilogrammes, making it the largest and heaviest land creature of the Americas. Bison originally evolved in central Asia and then migrated into America across the Bering Straits. The American bison are now the only remaining examples in the wild; those of the steppes have long since become extinct in the wild, although they have been bred in captivity and reintroduced to certain areas, such as the Bialowieza Forest in Poland.

When the Spanish arrived in 1541, about sixty million bison roamed the North American prairies, often in enormous herds of over a million animals, feeding on the northern plains during the summer and migrating over 500 kilometres to their winter feeding grounds in the south. Up until the end of the eighteenth century, their only predators were the Plains Indians, who depended on the bison for food, clothing and tools. Nevertheless, they maintained a balanced relationship with their prey, never killing too many

5

6

7

Current bison reserves

Bison distribution before 1860

8

The magnificent bison (**5,6,7**), the mature males of which stand 180 centimetres high at the shoulder and weigh up to 1,350 kilogrammes, was once found right across the North American prairies. From an estimated seventy-five million at the end of the fifteenth century, numbers dropped to only 540 individuals by 1889. Within the confines of a network of reserves, numbers have recovered somewhat today. The black korhaan (**8**) is a long-legged, largely terrestrial bird of the South African grasslands.

animals and using every single part of those that were killed.

Then the white settlers arrived. The massacre did not start until the 1830s but, by the early 1870s, it was estimated that some two-and-a-half million bison were being slaughtered annually. Furthermore, the building of the railway across the plains divided the bison herds in two and, by the end of the 1870s, the southern bison had been exterminated. By 1900 there were less than

1,000 wild bison left in existence, at which point it was realised that the species was in grave danger. All those remaining were herded into reserves, zoos and national parks, and their numbers have slowly built up again. Today there are between 30,000 and 50,000 of these magnificent animals in the United States' protected areas. As each bison requires about a hectare of tall-grass prairie, or about forty hectares of the less productive short-grass prairie, the available wild habitat

1

2

3

The Cape buffalo (**3**), reputed to be the most aggressive herbivore in the world, is a massive black beast standing up to 150 centimetres tall at the shoulder and weighing almost a tonne. The horns, which often span one-and-a-half metres, have heavy bases that meet across the brow and are employed in defence and in the ritual combat between males over females and territory. The nyala of southeastern Africa must drink regularly, living predominantly in swamps and along river banks; the females (**1**) lack the awl-like horns of the male nyala.

is already at full capacity, so never again will those vast, awe-inspiring herds of bison thunder across the prairie.

A similar fate befell the pronghorn, a mammal also known as the American goat antelope, because it displays characteristics of both these families. In the nineteenth century, their numbers on the American short-grass prairies were estimated at between fifty and a hundred million individuals. As they are considerably smaller than the bison, they have had to develop defences against wolves and other predators. Just as the bison is North America's largest land creature, so the pronghorn is the fastest , being capable of speeds of up to ninety kilometres per hour. Such speed, however, proved ineffective against man and, by the beginning of the twentieth century, only about 19,000 pronghorn remained. However, this story does have a somewhat happier ending: following stringent protection measures, pronghorn numbers are now approaching half a million.

A similar story can be told about the saiga antelope. In the eighteenth century, these ruminants were widespread throughout the Eurasian steppes, but they were hunted for their meat in such great numbers that,

by the beginning of the twentieth century, fewer than 1,000 individuals remained in the wild. Possibly as a result of the American example, man then realised that no other herbivore existed which was capable of converting the poor steppe pastures into meat. Thus hunting was prohibited and the remaining beasts protected and managed as a wild source of protein. Their recovery was even more dramatic than that of the pronghorn, for in the space of fifty years their numbers increased to more than two million. The Soviet Union today culls about ten per cent of that total for meat.

Przewalski's horse, which differs from domestic breeds by having an upright mane and no forelock, was an inhabitant of the Eurasian steppes. When it was first discovered, in the Gobi Desert in 1879, it was already rare, though it is thought to have occupied a vast area in prehistoric times. The numbers decreased dramatically between 1930 and 1950, both as a result of competition for pasture with the domestic animals of the nomadic tribes and of the continual political strife which afflicted the area. If Przewalski's horse still exists in the wild today, its numbers must be very low,

Brindled gnus are more commonly known as wildebeest, a word coined from the Afrikaans for 'wild cattle'. These long-legged antelopes are among the fastest herbivores on the savannas and undertake long annual migrations during the dry season in search of fresh pasturage along routes which were carved out centuries ago. Those from the Serengeti split into two groups, one heading west towards Lake Victoria and the other crossing the Mara River (**8**) and spending the summer further north.

Thompson's gazelles (**6**) sport thick black stripes along their flanks that break up their outline and so disorientate predators. Young springboks (**4**) are dependent on their mothers for six months; their sand-coloured coats help to camouflage them on the arid South African plains.

The female pronghorn of the North American prairies hides her young (**2**) separately in among the grasses, where they remain immobile, their pearly coats helping to conceal them from predators. The mother removes excrement from these hiding places so that its scent does not attract carnivores. Przewalski's horse (**5,7**), once found on the Eurasian steppes, is thought to be the only horse in the world never to have been domesticated, though none survive in the wild today. Fortunately, it breeds well in captivity and is in no immediate danger of extinction.

since the last definite sighting was in 1968 and extensive searches in recent years have failed to locate any trace of the animal. However, captive populations have been in existence for many decades – indeed, all the 660 individuals now in zoos were born in captivity – and eventually it is hoped some can be reintroduced to the wild, when the steppes will resound once again with the hoofbeats of Przewalski's horse.

In Africa there are two subspecies of white rhinoceros, the northern and the southern, whose populations probably became separated about two million years ago. At the turn of the century the southern race nearly became extinct in the wild as a result of excessive poaching to obtain rhino horn, highly valued as an aphrodisiac when powdered. The southern population recovered, however, and now numbers more than 3,500 rhinoceros, widely distributed throughout their original range. It would appear, however, that the northern race will not be so fortunate; there are only about fifty individuals in the wild today, despite the fact that at the beginning of the 1980s they numbered about 1,000. Their main strength is a breeding population of about fifteen animals at Garamba National Park in northern Zaire. Reintroduction to the wild is not at present a feasible option, since the world's zoos contain a mere thirteen specimens of the northern

Although yellow baboons (4) frequently take refuge in the trees overnight, they are essentially terrestrial creatures. Their long-fingered hands are equipped with an opposable thumb with which they transfer food – fruit, tubers and roots, fledglings and eggs and small reptiles – to their mouths. Mixed herds of zebra and wildebeest (6) are a common sight on the savanna as their grazing habits are complementary. The solitary black rhinoceros (8) may weigh some 2,000 kilogrammes and has a reputation for being bad-tempered and aggressive that is probably due in part to its extreme short-sightedness, which often causes it to over-react.

1

2

3

4

white rhinoceros.

Other African savanna species have also suffered at the hands of man. In the veld of southern Africa, many large herds of indigenous animals had been almost wiped out by the end of the nineteenth century and one, the quagga, had become extinct. A sort of half-finished zebra, striped on its head and neck only, the last wild quagga was killed in 1878, and the last captive beast only survived for a further five years. The white-tailed gnu is also extinct in the wild today, although it has fared better in captivity, there being about 500 head in reserves in southern Africa.

The only extensive natural grasslands of the world

5

6

7

Distribution of the Cobra

Predatory birds of the African savanna include the pale chanting goshawk (**1**), which has a melodious voice, in contrast to the harsh cries of most raptors, and the spotted eagle owl (**5**). Graceful chameleons (**3**) have cylindrical prehensile tails and laterally flattened bodies, as well as eyes which move independently of each other. The puff adder (**2**) relies on camouflage and ambush to trap its prey, whereas the black spitting cobra (**7**), the only African snake capable of projecting its poison over any distance, exhales violently and ejects a fine mist of venom and saliva at its victim's eyes, causing temporary blindness.

which remain more or less intact are the African tropical savannas. The East African savannas contain the highest concentrations of indigenous grazing animals in existence, probably because the land is not well-watered and is thus unsuitable either for domestic grazing stock or for cultivation. Because grassland communities are structurally much simpler than forest ones, it is possible, in the tropical savannas, to see exactly how the ecosystem works. We can follow the food chain from the lowest level, the grasses, right up to the big cats and birds of prey at the top.

Ultimately, all life depends on the sun as the source of energy used by plants for photosynthesis, the process whereby they manufacture carbohydrates in

8

order to grow. A herbivore is, by definition, any animal which obtains its food from the plant kingdom, although it need not necessarily feed on grass; some types of herbivore feed on the seeds or pollen of the savanna plants. As they feed, some species, such as hummingbirds, bats, rodents and even monkeys, help to pollinate the flowers or disperse the seeds. Other savanna animals are predominantly insectivorous, feeding on the thousands of grasshoppers, beetles and *Hymenoptera* – ants, bees and wasps. These animals include pangolins and aardvarks, whose diets consist almost exclusively of termites.

Within the savanna ecosystem it is the large herbivores which immediately attract attention, and none more so than the ruminants, amongst which there is considerable variation in size. The smallest grazers are species such as Thompson's gazelle and the impala, whilst the larger ones include the eland, wildebeest, roan antelope and topi; of these, the eland is the world's largest living antelope, the males weighing over 400 kilogrammes. Other animals, such as giraffes and gerenuks, have retained the browsing habits of their forest ancestors. The giraffe can in fact feed at any level from the ground up, using its extremely long, prehensile tongue to select only that foliage with the highest protein content. The non-ruminant herbivores include such veritable colossi as elephants and the rhinoceros, which demonstrates both types of eating

1

2

habit. The black rhinoceros has a prehensile lip and is predominantly a solitary, browsing creature, whilst the placid white rhinoceros lives in small groups and has a broad, square muzzle adapted to its grazing life-style.

These differences in the size and feeding habits of savanna herbivores enable the plant productivity of the savanna grasslands to be exploited to the full, but not to excess. Amongst the browsing animals, the gerenuk stands on its hind legs to nibble the lowest branches, the long neck of the giraffe enables it to feed on plants up to six metres above the ground, and the long trunk of the elephant permits it to reach shoots which are beyond the reach even of the giraffe. This utilisation of different parts of the vegetation also occurs amongst the grazing animals. The toughest grasses, for example, are eaten mainly by zebras, the only savanna grazers

with teeth in both jaws. Once this coarse vegetation has been removed, the leafy middle layers are eaten by the larger ruminants, such as wildebeest and topi, whilst the smaller gazelles feed mainly on the tender shoots at ground level. It is this segregation of feeding habits that allows mixed herds of over a million zebra, wildebeest and gazelles to roam the savannas together.

Within most of the extensive grassland zones, there are both seasonal and regional differences in climate, particularly rainfall, which the animals can exploit. A migratory life style provides them with fresh pastures all year round and also prevents the overgrazing of any one area. Perhaps the most spectacular migrations are those seen on the African savannas. At the beginning of May, the wildebeest herds of the Serengeti grasslands migrate to the northwest in response to the drying out

In the parched grasslands of the Etosha Pan, families of yellow mongooses (**8**) and Cape foxes (**4**) are sometimes found sharing underground burrow systems, the initial excavation of which was carried out by ground squirrels.

4

3

5

6

8

The impressive mane of the male lion (**3**) starts to develop at around two years, reaching about twenty-five centimetres in length at maturity. After the lionesses have made a kill, the feeding order is always males, females and then cubs (**5**), so that the young only reach maturity when food is plentiful.

7

of their winter habitat. They move in an orderly manner, forming columns across the plains, until they reach bottlenecks, such as river crossings, where many of them die in the ensuing crush, the survivors eventually reaching their destination, the Mara plains in Kenya. In November, when the rains have begun again in the Serengeti, they retrace their steps.

The non-migratory animals are largely the browsers, which are not dependent upon the seasonal growth of grasses, and include such species as the elephant, giraffe, warthog, rhinoceros and one of the smallest antelopes in the world, the dik-dik.

The elephant, possibly the ultimate herbivore, is the world's largest land mammal, weighing up to four tons, and has no enemies on the plains besides man. Most elephants, if they escape the hunter, will eventually die of starvation rather than as a direct result of old age. The reason for this is that elephants possess six molars in each half jaw, only one of which is in use at any given time. The others come into operation in a conveyor-belt fashion, pushing out the one above once this has been so worn down that it can no longer effectively grind up the tough vegetation on which the elephant feeds. When all six teeth have been used in this way, usually when the elephant is around sixty-five years old, the animal will slowly starve to death.

The tusks of the elephant are formed from the upper incisors, not from the canines, as in other tusked creatures; they are present in both males and females and have been known to weigh up to sixty kilogrammes apiece in a mature bull. They are a valuable tool for prising bark from trees for food and for digging waterholes. Unfortunately, their ivory is highly prized by man, and the poaching of the wild animals for their tusks has been responsible for the decimation of elephant herds throughout much of Africa.

Since an elephant's neck is so short that it is unable to reach the ground with its mouth, the trunk is invaluable for both drinking and dust-bathing. In essence, the trunk is a greatly extended muscular nose and upper lip, which the elephant can use so dexterously as to pluck the tender young shoots of the acacia from the top of the tree while avoiding the vicious thorns, which are the tree's defence against predators. The trunk can also be used as a snorkel when crossing rivers, which elephants do not by swimming, but by walking along the bottom.

The birds of the African savannas display a wide range of body size and shape and feeding preferences. Amongst the most numerous are the weaverbirds, which adorn the acacia with intricately constructed nests, like so many hanging baskets, designed to keep their young safe from predators. The red-billed quelea and masked weaverbirds are two of the most common

White-fronted bee-eaters (**2**) are colonial nesters, having perhaps the most complex social system of any bird. A breeding pair has up to three 'helpers' to assist with nest excavation, incubation of the eggs and feeding the chicks, so increasing the chances of survival in each brood.

2

3

1

4

With skins up to three centimetres thick, African elephants (**3**) would have problems with heat loss were it not for their large ears, which increase the surface area to volume ratio, and their habit of mud-bathing, which cools the body through evaporation (**5**).

A male ostrich may serve several females, each of which lays her eggs in a communal nest, but only one pair is responsible for rearing the young (**7,8**). Secretary birds (**4**) pair for life and the nest, some two metres in diameter, is constructed in the top of a thorny acacia. The vulturine guinea fowl (**6**), named for the patches of naked blue skin on its head, is a gregarious bird that prefers to run from danger rather than fly. Male red bishops (**1**), relatives of the weaverbirds, are handsome birds. Their fledglings form huge nomadic flocks which cause great damage to sorghum crops.

species, but there are many more, each with different ideas on nest-building. These birds maintain their populations in equilibrium with their environment. If the rains fail and the grasses do not produce sufficiently long leaves and stems to enable the birds to build their nests, they do not breed, thus preventing overpopulation at a critical time.

Ground-dwelling birds, not surprisingly, are also numerous in the open plains. Many of these have become so well-adapted to life on the ground that they are now loth to take to the air. The most extreme example of this is the ostrich, which has completely lost the ability to fly. Nature has compensated the ostrich in other ways, however, providing it with a pair of long, enormously powerful legs on which it can cover great

distances quickly, at speeds of up to seventy kilometres per hour, and effortlessly. The ostrich is one of the largest birds in existence, weighing about 135 kilogrammes and growing to well over two metres in height.

Other African ground-dwelling birds include black hornbills, several species of sandgrouse, crowned guinea fowl and ground-nesting crested cranes. One of the most striking examples is the secretary bird, a member of the falcon family, which has greatly elongated legs and strides through the tall grass searching for lizards or small mammals which it dispatches with its strong, curved talons. Despite spending much of its time on the ground, the secretary bird has retained the power of flight and nests in the tops of acacia trees. The kori bustard, at about twenty-three kilogrammes the heaviest of all flying birds, is another inhabitant of the savanna grasslands.

Although they live by exploiting the grasses of the savannas, herbivores also fulfil a vital role in maintaining the vigour of the grasses, both by grazing and fertilising and by eating or trampling down the woody shoots which threaten to invade the plains. The predators in turn thin out the herbivores, thus preventing overgrazing of the grasslands. The most successful carnivores of the savanna catch their prey in many different ways. Hunting dogs, for example, work as a team, characteristically attacking from the rear, whilst lions always go for the throat. The cheetah, the fastest animal on earth, is supremely adapted to the open plains. Where else could it reach speeds of up to 120 kilometres per hour unhampered by vegetation? Although faster than any of its victims, the cheetah will not always succeed in making a kill, since it can only maintain this velocity over short distances. One feature which prevents predators having to compete for the same food supply is that each generally attacks victims of approximately its own size.

After the predators have eaten their fill, it is the turn of the scavengers – the vultures, marabou storks, white-necked ravens and hyenas. The vultures spend most of their time soaring over the plain, barely flapping their wings, so as to save energy, as they glide on the warm updraughts of air. Once a vulture has spotted an animal carcass it will circle down rapidly to feed. By dint of keeping one eye on the ground and one on their companions in the skies, all the other vultures in the area will make sure they are invited to the feast. The unattractive bald heads and necks of many vulture species actually serve a purpose. Species such as the white-headed and black-backed vultures delve deep into the carcass when feeding and, if they had feathers

African bat-eared foxes (**1**) possess enormous ears which act as heat exchangers and are also invaluable for detecting the slightest movement of a small mammal or lizard during their nocturnal hunting forays. When the foxes are very young their ears are floppy, but they soon become erect.

1

2

3

4

on their heads and necks, would not be able to rid them of rotting meat. The Egyptian vulture, on the other hand, is much more comely, with an all-over plumage of white feathers; it has adopted a different approach to scavenging, waiting for scraps to be tossed aside by other feeders.

The spotted hyena is also well-adapted to its scavenging life style, being armed with powerful jaws to crack open even the largest bones to extract the tender marrow from inside. Lesser creatures also play their part in tidying the savannas; carrion flies and beetles lay their eggs in the dead animal. Such is the efficiency of the scavengers that only hours after they descend on a carcass little more than bare bones remain.

The food chain of an ecosystem may appear to

5

Yellow baboons (**3**) fear the night as their chief predator – the leopard – is then on the prowl. As a result they will spend the hours of darkness huddled in the tallest branches of a tree, only descending to feed and drink when daylight returns.

Lions (**6**) must have regular access to water and often take fifteen minutes to quench their thirst. Young lions (**4**) do not take an active part in hunting expeditions until they are about a year old and have permanent teeth. Cheetahs (**7**) are rarely found in the same areas as lions and leopards, although they are not in direct competition. Spotted hyenas (**5**), with their robust forequarters, weak hind legs and powerful jaws, are considered to be the archetypal carrion-eaters. During their nocturnal hunting trips, however, packs of hyenas make their own kills, scavenging only during daylight hours. The black-backed jackal, seen here at a wildebeest carcass surrounded by vultures (**10**), is thought of primarily as a scavenger, yet, hunting in pairs, jackals often bring down gazelles. The lappet-faced vulture (**2**), in using its powerful beak to feed, opens up animal carcasses to less able scavengers such as the white-backed vulture (**8**).

6

8

7

9

terminate at the top of the pyramid with the carnivores and scavengers, but it is in fact cyclical, since waste products formed at any level, whether faeces, dead plant material or uneaten animal remains, are decomposed and returned to the soil. The majority of the creatures responsible for this process, the decomposers, are too small to be visible to the naked eye. For the most part they live in the soil, tiny single-celled animals, such as bacteria and protozoa, as well as fungi. A somewhat larger help in tidying up the savannas is the dung beetle. Only a few days after the great animal migrations, there remains not a trace of the animals' passing. The dung beetles roll the faeces into balls and bury them underground, not for any altruistic reason, but to provide a food supply for their developing larvae.

10

Africa's answer to the Eurasian wild boar, the omnivorous warthog is not one of nature's most handsome creatures. It has four lumpy warts on its face, two below the eyes and two above the snout, and its four large canine tusks curve outwards and upwards (**9**).

A pack of spotted hyenas reclining in the early morning sun in Botswana were not too sleepy to notice the appearance of a large male warthog in the distance. Having pursued the unfortunate beast and brought it to its knees (**1**), the hyenas were able to devour about half of the carcass (**2**) before a group of lionesses appeared on the scene. The scavenging lionesses drove off the hyenas and proceeded to demolish the remainder of the warthog (**3**), joined within a few moments by about a dozen hungry cubs (**5**,**6**,**7**). A group of black-backed jackals had been alerted by the warthog's death cries and hovered on the periphery of the feeding lions alongside the disappointed hyenas (**4**). Discord was rife even among the lionesses and their cubs, late arrivals being warned off the carcass by those in occupation (**8**) and one lioness attempting to pilfer the warthog's head from the general melee (**9**).

7

8

9

10

11

It is clear from these pictures that the hyena's reputation for living off scraps left over from the hunting exploits of more 'noble' beasts such as the lion is rather unjustified, as in many cases the hyenas make the kill, only to be displaced by larger predators and reduced to squabbling over the bones.

The great grasslands of the world are widely separated geographically, but it is quite astonishing how the typical fauna of each region has adapted in similar ways to fill the available niches. Thus almost every one of Africa's savanna creatures has an ecological equivalent, although not necessarily a direct relative, elsewhere in the world. Perhaps the greatest similarities occur between the grassland animals of Africa and those of South America, probably due to the fact that these continents are both in the southern hemisphere and are subject to more or less the same climate.

The scaly pangolin and strange-looking aardvark of Africa are thus mirrored in the twenty-one species of armadillo and the giant anteater of South America.

The South American buff-necked ibis (**1**), like all members of the ibis family, has a long, downcurved bill and feeds mainly on small fish, amphibians and invertebrates. Its nest is a bulky affair, usually a stick platform constructed in the treetops.

With fifty-eight genera and 221 species, the ovenbird family is the most diverse in the avian world. Rufous ovenbirds (**5**) occur throughout the open plains of southern Brazil, Paraguay and Argentina, where, as their name suggests, they build dome-shaped nests of mud reinforced with grass. Ground-dwelling tinamous, found from Mexico to Patagonia, are plump, dull-coloured birds whose instinctive reaction to danger is to freeze, taking to the air only as a last resort. The elegant crested tinamou (**7**) is one of the most distinctive of its kind.

Although these creatures do not all belong to the same family, each feeds almost exclusively on ants and termites, having strong claws with which to extract the insects from their mounds and a long, sticky tongue with which to scoop them up. In the giant anteater, the tongue may be sixty centimetres long and bears minute, backwards-pointing spines to prevent the ants escaping. It has no teeth, but the stomach is very muscular and contains small amounts of gravel with which to pulverise its contents. In order to survive, this bizarre creature must consume about 30,000 ants or termites per day.

In the same way, ostriches are analogous to the rheas of the New World; in both cases the males are left to look after the eggs and rear the young – a true case of sexual equality among birds! The secretary bird has adopted an identical way of life to the seriema, although the former is a type of falcon and the latter belongs to the crane family, and the South American screamers are also long-legged plains birds, in this case related to ducks and geese. The falcons of the savannas are represented in the Americas by a related group, the caracaras, although these are thought to derive from more ancient stock. African sandgrouse have their South American counterparts in the partridge-like tinamous, and the weaverbirds of the acacias correspond to the ovenbirds, nest builders with a somewhat different

6

The most distinctive grassland mammals of South America are undoubtedly the armadillos, the nine-banded armadillo (**3**) occupying a vast range extending from the southern United States to the Argentinian pampas. Among the more primitive species is the hairy armadillo (**2**), thought to be a miniature replica of the extinct glyptodonts. Guanacos (**4**) often graze alongside rheas, the sharp sight of the birds complementing the excellent senses of smell and hearing of these New World camelids in recognising danger. Maned wolves (**6,8**), with their long slender legs and keen vision, are capable of short sprints, but they prefer to feed on small vertebrates, insects and even fruit.

7

architectural style. These small birds construct their mud and grass nests, resembling tiny ovens, on low branches or fence posts. The nests are baked rock hard by the sun and have a dividing wall just inside the entrance to prevent unwelcome paws or snouts from reaching the eggs or chicks.

The dog family is perhaps better represented in South America than in Africa. The South American pampas fox is more or less analogous to the African jackal, whilst the maned wolf possibly fulfils the same role as the cheetah, since it has inordinately long legs, giving the body equal proportions in both length and height. Like the cheetah, its stamina is poor and for this reason it seldom chooses to hunt large animals, instead preying mainly on small burrowing mammals such as viscachas and tucutucus, and on surface-dwelling creatures such as the pampas guinea pig. In the cat family the South American jaguar can be compared to the African leopard in terms of its agility in the trees, although today most of them have retreated into the jungles before the ever-advancing hordes of man.

The only animals missing from the pampas and llanos are the huge herds of antelopes and deer, which were late reaching Latin America due to the absence of a land bridge with the north. Pampas deer, marsh deer, whitetails and red brockets are virtually their only

8

Flightless emus (**8**) and red kangaroos (**6**) are true creatures of the Australian steppes and savannas, but the grey kangaroo (**2,5**) is found in habitats with more vegetation and higher rainfall. The male emu rears the chicks for some eighteen months.

representatives; instead, the characteristic herbivore of the llanos is the capybara, the world's largest living rodent, which is about the size of a domestic pig. Capybaras are perfectly adapted to cope with the annual flooding of the llanos, having their eyes, ears and nostrils all set on top of the head, and webbed feet.

Hares too, so typical of other grasslands, are not found in South America, so a type of rodent, the mara, or Patagonian cavy, has evolved to fill this niche. Although resembling a hare in some respects, the mara has enormously long hind legs, with which it can execute leaps of two metres with ease. Because it is a burrowing animal, it has robust claws with which to excavate its tunnels, and eyes positioned near the top of its head to allow it to scan the surrounding plain without leaving the safety of its underground retreat.

Moving across to Australia, it is interesting to note that here there is only one type of large herbivore, the kangaroo, although there are many species within this group. The red kangaroo has the speed and economic gait typical of many African grazers; normally it prefers to travel at about twenty kilometres per hour, but it can increase this to sixty-five kilometres per hour in order to escape danger. The kangaroo's bounding action, in contrast to a four-legged gait, actually assists breathing, by mechanically squeezing the air out of the lungs on impact, whilst the tendons in the hind legs act as springs, storing the energy imparted by one bounce for use in the next. Although the kangaroo cannot be said to resemble any creature outside Australia, the emu,

another large flightless bird, bears a striking resemblance to both the African ostrich and the South American rhea.

In Australia, as in Africa and South America, millions of termites construct mounds ranging from narrow pinnacles to broad, squat domes and, as one would expect, a creature has evolved to exploit these social insects – the echidna. Echidnas belong to a group of primitive egg-laying mammals called the

Bruijn's wallaby (**4**) is one of a range of marsupials which has spread from Australasia into Indonesia.

monotremes, of which there are only three living representatives in the world today: the duck-billed platypus and two species of echidna, one of which occurs only in the jungles of Papua New Guinea. Like the other ant-eating creatures of the world, the echidna, also known as the spiny anteater, has strong claws with which to break open the nests or mounds, but it is also coated with a fearsome array of sharp spines. Like some species of armadillo, it can roll itself into a ball when danger threatens.

The small burrowing mammals are represented in Australia by nineteen species of marsupial bandicoot and a wide variety of hopping mice, whilst on the steppes of central Asia and eastern Europe the

6

The brush-tailed possum, or Adelaide chinchilla (3), is a nocturnal marsupial native to Australia, whereas the dingo (7), thought to have arrived around 5,000 years ago, spread very quickly and will probably be ultimately responsible for the demise of the thylacine.

7

8

9

10

11

Termite mounds (9) are constructed from a mixture of clay and termite saliva; their internal temperature is a constant 30°C and atmospheric humidity close on a hundred per cent. Each highly-organised colony contains a single grotesque queen (10) and a smaller king, together with well-defined worker and soldier castes (11). The white cockatoo (12) is an Australasian species measuring fifty centimetres from head to tail. (1) An Indonesian blue-throated bee-eater partakes of a dragonfly.

predominant tunnelling creature is the European suslik. These rotund little mammals hibernate for half the year and spend the other half feeding themselves up and burying caches of seed underground in preparation for the next period of dormancy. Such has been man's interference with the natural ecosystem of the steppe that it is the burrowing creatures, which can stay out of harm's way, which are by far the most numerous species today. For example, within an area of about three hectares there may be up to 12,000 suslik burrow entrances. Other burrowing denizens of the steppe include lesser mole rats, which use their teeth to excavate their burrows, steppe lemmings, black-bellied hamsters and Bobac marmots.

The steppes, although they once thronged with wild cattle and horses, are now the lonely domain of a single species of herbivore – the saiga antelope. Like its African counterparts, the saiga antelope is highly nomadic, gathering in the more humid river valleys as the plains dry out in August and later travelling south 12 in herds of up to 100,000 to escape the bitter winters of the northern steppes. At times they have been known to travel some 240 kilometres per week, often in large herds. Despite these measures to escape the harsh winters, there are times when many thousands die of hypothermia and starvation. By way of compensation, the females are extraordinarily fertile, being able to mate for the first time at only four months, before they are fully grown, and generally producing twins. In this way the saiga population can be built up again rapidly following a winter crash.

Living on the dusty steppes, the saiga antelope has developed a bizarre facial appearance, having bulbous eyes and a distinctly Roman nose which ends in a stubby, flexible trunk and has wide, circular nostrils.

1

2

3

4

5

The cheetah (3), once one of the most widespread predators of Asia, has all but vanished from this region due to intense persecution and a reduction in prey availability. For similar reasons, the Asiatic lion (4,5) is one of the rarest carnivores in the world today, found only in the Gir Forest Reserve on the Kathiawar peninsula. The leopard (1), however, because it can survive in a more varied range of habitats, is still abundant in Asia, the steppe animals being paler and more slender than their forest counterparts. Likewise, Asiatic elephants (6), although suffering from the destruction of their forest environment, are still quite common.

Inside this large nose there are a number of chambers lined with mucous glands; scientists are of the opinion that these serve to filter dust from the air as the antelope runs, which it does with its head close to the ground, at speeds of up to eighty kilometres per hour.

The ground-nesting steppe eagle is one of the largest birds of this region, but it is a mere dwarf compared to the stately great bustard that stalks the plains. Bustards are true open grassland species, found throughout the world. Perhaps the rarest of these today is the great Indian bustard, which lives in the arid Rajasthan region, where the vegetation is low enough for the bird to see clearly in all directions. Its wingspan often exceeds two metres, although, like most large grassland birds it prefers not to take to the air. In keeping with its more terrestrial lifestyle the backwards-pointing toes have been lost, as in the ostrich and other ratites, to facilitate an energy-efficient gait. As with all bustards, the courtship display of the male is nothing short of magnificent: the wings and tail feathers are spread out and the neck pouch swells to emit deep, booming calls, which attract the attention of females

The slender caracal (2) is also a common Asian carnivore, not competing with similar-sized predators for food since it has a predilection for birds, which it is expert at catching on the wing.

6

7

A rigid class system exists within baboon troops, with young and adolescent individuals (**10**) at the lowest levels and the upper echelons occupied by a number of dominant males.

8

9

10

11

The Asian great thick-knee (**11**) is a plains-dwelling wader very similar in appearance to the stone curlew. Avian steppe predators range from rufous-backed shrikes (**7**), aggressive birds with slightly hooked beaks that are renowned for impaling their victims on thorns in 'larders', to massive eagles. The Ceylon hawk eagle (**8**), a subspecies of the crested hawk eagle, is a powerful hunter on grassland and in forest, while the steppe eagle (**9**) is a dark variant of the tawny eagle found throughout the Eurasian wooded plains.

up to a kilometre distant. In order to avoid predators, the chicks are able to walk from the moment they hatch.

Predators are few and far between in the Asian grasslands today, however. The marbled polecat feeds predominantly on the burrowing rodents, digging them out of their tunnels with its powerful claws, but any larger carnivores have all but disappeared. The Asiatic cheetah is now rarely seen in the wild and the counterpart of the African lion, the Asiatic race, is now confined to the Gir Forest of India. Here, in about 1,300 square kilometres of dry, open terrain, their final retreat since about 1940, live the last 200 Asiatic lions. They differ little from their African counterparts, although they are somewhat smaller and have a less pronounced mane.

It is also possible to see similarities between the animals of the North American prairies and those of other grassland regions. Possibly because of its less favourable climate, only two large herbivores, the bison and the pronghorn, have ever evolved in the temperate prairie. The swift, long-legged pronghorn antelope has a much wider windpipe than other antelopes so as to be able to breathe sufficiently whilst

travelling at speed; it also runs with its mouth open and tongue hanging out. Unlike many mammals it runs in a straight line, thus outdistancing predators such as wolves and coyotes.

The South American viscachas and the susliks of the steppes have their biological equivalent in North America in the prairie dog. These creatures, in reality colonial squirrels, have perfected the art of burrowing and construct huge underground cities; one such complex was estimated to contain some 400 million individuals! Because they feed on the crops and resown pastures that now cover most of the prairie lands, they were subjected to a well-organised programme aimed at their extermination in the first half of this century. The prairie dog managed to survive, but the reduction

Crab spiders (**1**), so called because of their sideways manner of walking, adopt the colour of a particular flower and ambush their prey from within it. The lubber grasshopper (**5**) is a denizen of the tall-grass prairies capable of consuming vast quantities of vegetation. Sagebrush provides both food and cover for the sage grouse (**2**), largest of all New World grouse; the splendid male sports orange inflatable air sacs used in courtship displays as well as an elegant white neck-ruff. Meadow larks (**3**) bear the typical drab plumage of many ground-nesting birds of the North American prairies.

grouse, which have a courtship display, complete with 'booming', similar to that of the bustards.

When ancestral man emerged from the forests of Africa about two million years ago, he found himself well-adapted to life on the plains. A bipedal stance was a considerable advantage in allowing him to see over great distances and the position of his eyes permitted a wide field of vision. His opposing thumbs and improved hand-eye coordination made it possible for

6

7

Ruby-throated hummingbirds (**7**) stray further north than any other hummingbird, breeding in southwest Alaska, although they winter in Central America.

8

9

Prairie dogs are in fact members of the squirrel family, very territorial and with a highly developed social organisation. The black-tailed prairie dog (**9**) is found mainly in the eastern prairies. The devastation of the prairie dog population has rendered the black-footed ferret (**6**) the rarest of North American mammals. Coyotes (**8**) also have a predilection for prairie dogs, but are much more versatile in their feeding habits, eating anything from insects to domestic animals and carrion; for this reason they still abound on the American plains. The Mexican wolf (**4**) is another highly opportunistic predator, feeding on whatever is available.

in its numbers had serious consequences for some of its predators, such as the American badger, which is probably the equivalent of the marbled polecat in the steppe, and the black-footed ferret. The latter has been almost wiped out: there are only ten known individuals left on the Wyoming prairies and the species appears destined to become extinct in the near future.

The coyote is the northern counterpart of the pampas fox and jackal, and the caracara of South America is replaced on the prairie by the red-tailed hawk. But the anteaters, aardvarks and echidnas have no opposite number on either steppe or prairie because there are so few species of ants and termites in the temperate grasslands of the northern hemisphere, and most of these do not form large colonies. The niche of the ground-dwelling bird, however, is filled in North America by several species known collectively as prairie

him to throw rocks, and later spears, at his prey. He also discovered the use of fire as a hunting tool, to flush his prey from its cover. Just as ancient man helped to shape and maintain the primeval grasslands, so his descendants have carried on the tradition. The first grain-growing experiments started some 8,000 to 10,000 years ago, and the American grasslands alone now support annual crops worth nearly 150 billion dollars.

Unfortunately, if you wish to see natural grassland ecosystems intact today, a trip to Africa is your only alternative. In almost every other part of the world, man has decimated the wild herds and converted the grasslands to his own use, with little regard for the ecological consequences of his actions. Where vast plains once teemed with life, today there is only dust swirling in the wind.

· CHAPTER 2 ·

ARID LANDS

Natural arid lands cover about one third of the earth's continental surface. From the bitter cold of the barren, stony plains of Mongolia's Gobi Desert to the scorching sands of the Namib Desert in southwest Africa, they all have one thing in common: a lack of water. Drought may be a seasonal occurrence, as it is in the semi-arid regions of the Mediterranean, southern California and Chile, or it may last several decades, as it does in the eastern Sahara and central Australia. Yet these are not lifeless wastelands, for all their desolate aspect. Over the course of millennia, since the deserts of the world were first formed, animals and plants have been perfecting their survival strategies in this uncompromising domain, where water is scarce and temperatures are extreme. Out of these survival strategies have evolved some of the most exquisite and bizarre creatures that grace our planet today.

Right: fingers of organ-pipe cactus, Arizona, (top) a lanner
falcon and (above) a sandgrouse chick.

True deserts are distinguished from arid and semi-arid lands by their extreme lack of water. At worst, the amount they receive, whether as rain, dew, mist or even snow, is less than 100 millimetres per year, whilst arid lands may receive up to 250 millimetres and semi-arid about 600 millimetres. The percentage of this then lost by evaporation is also significant, since, where the sun beats down relentlessly, evaporation further decreases the amount of water available to living things. The extent of these parched regions today is quite phenomenal. One third of the earth's landmass is classified as semi-arid or arid and about half of this is true desert.

Almost all of the world's true deserts lie between the latitudes of 15° and 30° on each side of the equator, coinciding more or less with the tropics of Capricorn and Cancer. The world's arid lands display an amazing degree of symmetry and, for every desert region in the northern hemisphere, there is a corresponding arid zone to the south. Thus the Sahara in the north is mirrored in the southern hemisphere by the Kalahari and Namib deserts; the Mojave and Sonoran deserts of the American southwest are echoed by the Atacama and Sechura deserts of Chile and Peru; and the enormous desert zone of central Asia is balanced in the south by the arid lands of Australia.

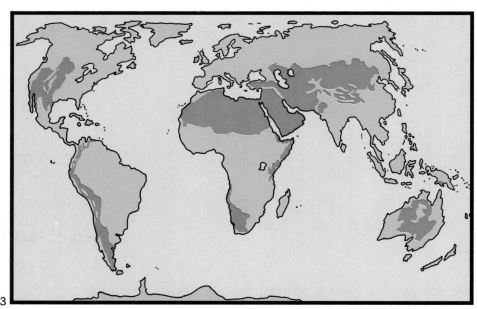

Europe is the only continent which has no true desert region, although arid and semi-arid lands are widespread, especially in the Mediterranean Basin, around the Black and Caspian seas and in parts of the Ukraine and North Caucasus. Elsewhere, some countries, particularly those in the region of the world's largest desert, the Sahara, in North Africa, consist almost entirely of arid lands. Egypt, for example, is ninety-six per cent desert, two thirds of which comprises the Libyan, or Western, Desert. Here, in an area

known as the Great Sand Sea, the sun is hot enough to evaporate more than 200 times the actual rainfall. As it is, rain is so rare an occurrence that whole generations of mankind are unable to describe it.

In southwest Africa the Kalahari covers about eighty-five per cent of Botswana, whilst the adjacent state of Namibia is named after its extensive arid coastal strip, the Namib Desert. Much of the Arabian peninsula is covered by the Rub' al-Khali and Nafud deserts and central Asia boasts an enormous desert

The world's arid lands (3) are distributed more or less symmetrically either side of the equator. (5) Cardon cactus silhouettes evoke the atmosphere of the Californian deserts.

5

vast arid area of the southern hemisphere covers much of the interior of Australia and is divided between the Great Victoria, Gibson, Great Sandy, Simpson and Sturt Stony deserts.

There are various reasons why some parts of the world should be so deprived of rainfall as to become deserts or arid lands. Firstly, the global circulation of air masses is such that two belts of dry, high-pressure air are always present over the tropics of Cancer and Capricorn, within whose latitudes most of the world's

Domestic animals (**6**) are one of the greatest agents of desertification in Africa following the abandonment of the traditional nomadic way of life.

6

7

The Arabian oryx (**1**), extinct in the wild by 1972, has since been successfully reintroduced to its natural habitat. Among the indigenous peoples of the Arabian deserts oryx blood is thought to counteract the effects of snakebite. In the heart of Africa's arid lands even a pride of mighty lions (**7**) defers to a herd of elephants on the move. (**2**) Two cold-blooded reptiles – the savanna monitor and the spitting cobra – bask in the early African sun so as to raise their body temperatures sufficiently to be able to hunt. Spitting cobras aim a mixture of saliva and venom at the eyes of their intended victims, causing temporary blindness. The cardinal (**4**) of the Sonoran Desert is named for its vivid crimson crest, which resembles the mitre of the appropriate Roman Catholic Church official.

belt, stretching from Pakistan's Baluchistan plateau northwards to the Turkestan Desert of the Soviet Union and eastwards into China's Tarim Basin and Mongolia's Gobi Desert. In the New World, the peninsula of Baja California is almost entirely arid, connecting to the north with the Great Basin Desert and to the east with the Mojave, Sonoran and Chihuahua deserts. The Atacama Desert comprises a large proportion of Chile, whilst neighbouring Argentina contains the Monte and Patagonian deserts. The other

true deserts occur. Secondly, in some areas, the global circulation of the seas results in cold coastal currents, which chill the air above, thus reducing its moisture-carrying capacity. The Atacama Desert of Chile owes its existence to this effect. The Humboldt Current, which flows up from Antarctica, cools the prevailing winds and reduces their moisture-carrying capacity, causing dense fogs and sea mists for much of the year. On reaching land, however, the air is reheated and able to carry more moisture, which it absorbs from the continental air-masses. In between, rain rarely falls on the coastal strip of the Atacama Desert itself and conditions there are so severe that in places there is no life at all.

Coastal mountain ranges give rise to a similar effect, causing moisture-laden air sweeping in from the oceans to rise, cool and discharge that moisture as rain over the mountains. Little moisture then remains for the area immediately behind the mountain range, which is said to lie in the rain shadow. In the southern half of Latin America, where the towering ridge of the

Andes runs along the western coast and the winds emanate from the South Pacific, the arid Patagonian plains cover a huge area immediately to the east of the mountains, lying in just such a rain shadow.

Another reason for the existence of deserts is based on much the same principle. The central regions of large continental landmasses, such as Eurasia, are often thousands of kilometres from the sea and any winds which reach these regions have long since discharged the moisture they were carrying. The vast Asian deserts of the Gobi and the Tarim Basin demonstrate this phenomenon, as does the interior of

The range of the red kangaroo (**2**) extends unbroken across the whole of Australia's semi-arid interior, some two-and-a-half million square kilometres, although numbers vary according to the availability of water, grazing and shelter from the sun.

The Sonoran Desert of the American southwest is home to such bizarre creatures as the chuckwalla (**1**), which may reach half a metre in length and is strictly vegetarian. This iguanid lizard escapes the clutches of avian predators by wedging itself into a crevice and inflating its lungs. The sidewinder rattlesnake (**4**) is also a denizen of the Sonoran Desert. It moves sideways in a series of S-shaped flicks of the body for minimal contact with the scorching sands, a method also adopted by the sidewinding adder of the Namib. In the absence of trees, the Sonoran cactus wren (**5**) will nest in the spiny upper regions of a giant cactus or yucca. The strong, lance-shaped horns of the gemsbok are an efficacious defence against even large predators, which is why the presence of a male lion (**3**) at an Etosha waterhole does not unduly worry a group of large, powerful oryx.

The century plant (**6**) of the American arid lands produces only a few leaves each year, gradually building up enormous food reserves before it flowers and dies. Although it is called the century plant, flowering usually occurs after fifty to sixty years.

even a relatively small landmass, such as the Iberian Peninsula, where summer droughts are a regular climatic feature.

Most of the world's deserts were formed at the same time. During the ancient tectonic movements in the earth's crust, numerous mountain ranges were forced into existence, causing global changes in the circulation of the world's air masses. These changes in turn brought about effects such as have been outlined above.

There is considerable evidence that many of the world's deserts were once fertile plains. Wall paintings in the Sahara depict animals such as the hippopotamus, a creature never found far from water, and the discovery of Stone Age axes in the Namib Desert seems to indicate that this region was not always as barren as it is today. It is thought that the Asian desert basin was once a vast inland sea, whilst the centre of Australia used to house an extensive lake and river system.

The world's arid lands can be divided into two distinct types: hot or cold. Generally, the further from the equator the desert is situated, or the higher its altitude, the colder it will be, both at night and during the winter.

The Sahara, lying flush with the Tropic of Cancer and extending across northern Africa from the Atlantic

Ocean to the Red Sea in a belt 2,000 kilometres wide, is essentially a hot desert. However, the lack of cloud cover over the area causes great variations in temperature. During the day temperatures in the sun may exceed 70°C and the highest shade temperatures in the world have also been recorded here. By contrast, night-time temperatures drop rapidly because there is no layer of cloud to retain the daytime heat. Finally, in only a few areas of the Sahara is there more than 100mm of rain per year and this sometimes falls in a single storm.

Death Valley, lying between the Great Basin and Mojave deserts, forms the lowest-lying land in North America at eighty-two metres below sea level and rainfall is often as little as fifty milimetres per year. It is

the north, thus exposing the area to icy air sweeping down from the Arctic.

The Namib Desert forms a long, narrow strip along Africa's southwest coast, stretching for over 2,100 kilometres from the Olifants River in Cape Province northwards into Angola. There is very little rainfall and parts of this desert are considered to be some of the most desolate places on earth, supporting no form of life. Other parts of the Namib Desert, however, are affected by the cold Benguela Current, which streams northwards along the coast from Antarctica and causes the moist, relatively warm air masses moving across the Atlantic Ocean to cool rapidly. The effect is identical to that already described in the case of Chile's Atacama Desert. Thick mists sweep across the dunes

The barrel cactus (**1**) of California is ideally shaped to withstand drought, possessing the minimum surface area over which water can be lost and the maximum bulk in which to store it. The Sonoran fish-hook cactus (**3**) bears a complicated arrangement of spines which both protects the plant from grazing animals and cuts down air movement around the plant to decrease water loss. The massive organ pipe cactus (**4**) compensates for the lack of regular rainfall in Arizona's arid lands through its tremendous capacity to store water when it is available. Its pleated stem allows the cactus to expand to almost twice its normal girth. Like all cacti, the strawberry hedgehog cactus (**2**) has brilliantly-coloured flowers.

2

here that the hottest ever air temperatures in North America, close to 57°C, have been recorded, yet at night the temperature is often near freezing. Despite these extremes, there are more than 600 thriving plant species and scientists have identified over 100 birds and about forty mammals.

By contrast, in Mongolia's Gobi Desert night-time temperatures of -40°C have been recorded during the winter, which is both severe and prolonged. The Aral Sea, to the north of the Kara Kum Desert, freezes over every year for four or five months and, to the east of this region, much of central Asia is covered with snow for long periods. The main reason for these periods of severe weather is the absence of mountain ranges to

1

3

4

5

The ground cuckoo (**6**) of the southwest American deserts is probably better known as the roadrunner because of its ability to reach speeds of around forty kilometres per hour. Its prey – small mammals, birds and reptiles – is steadily pursued to the point of exhaustion before being despatched by repeated blows of the roadrunner's powerful bill. A chainfruit cactus is an ideal nest site (**7**), providing the young roadrunners with shelter from the sun and protection from predators. The saguaro cactus (**8**) of the Sonoran Desert provides food and nest sites for many animals. Gila woodpeckers excavate nest holes in it, later used by elf owls, insects and bats feed from the flowers and its fruit is eaten by white-winged doves and other birds.
(**5**) The arid Valley of Desolation, where rocky outcrops tower above Graaff-Reinet and the Karoo, South Africa.

6

7

on about sixty nights of the year and a unique community of animals and plants has evolved to take advantage of these freak conditions in what is otherwise a parched desert region.

It is a common misconception that all deserts consist of wave after wave of sand dunes. In fact, many different types of terrain exist and it is simply their lack of water which forms a common denominator. The Sahara, for example, comprises three main types of terrain: hamadas, which are stone plateaus up to 3,300 metres high; regs, which are extensive areas of silt, gravel and stone laid down by previous floodwaters; and ergs, seas of sand dunes which conform to the popular idea of desert scenery, although they actually cover less than ten per cent of the total area of the Sahara. On the other hand, the Kara Kum Desert in Asia covers almost 500,000 square kilometres, three quarters of which consists of loose sand. Here the crescent-shaped dunes, the 'horns' of which point downwind, are known as barchans. Like most sand

dunes, they are not static, being continually driven before the wind.

In Australia there are five desert types, from the clay-pan formations of the Sturt Stony Desert to the largest sand-ridge desert in the world, the Simpson, where the dunes, which can be up to thirty-five metres high, stretch for over 120 kilometres in long, parallel waves. Even these are mere ripples compared to the 300-metre-high giants of the southern Namib Desert, which are the largest known sand dunes in the world. Arid lands are thus characterised by lack of water, extreme variations in temperature and, in many cases, a loose, shifting surface. These are obviously not favourable conditions for the successful survival of wildlife, and the process of adaptation that desert plants and animals have undergone has resulted in some of the world's most distinctive species.

Water is the most restricting factor and those plants, known as xerophytic species, which are typical of arid regions, have evolved countless different methods

8

of making the most of what little moisture there is. A plant's root system is all-important for extracting water and nutrients from the soil and two distinctly different methods have been evolved, to do this.

Even in those areas where rainfall is more frequent, often falling as one or two showers per year, the water cannot penetrate far into the soil before being lost through evaporation. In order to exploit this transient water supply, many plants have developed extensive but shallow root systems. The sand acacia of the Kara Kum Desert, for example, has a root system which extends up to fifteen metres away from the plant in any direction, while the camel-thorn of the same region may have roots up to thirty metres long. But perhaps the best example of this principle is the creosote bush of America's southwest, whose shallow roots are designed to utilise the film of dew which forms daily on the soil particles just below the surface. These bushes are widely spaced, at more or less regular intervals. As no seeds can germinate in the intervening soil, from which the bushes' roots have extracted all trace of moisture, they reproduce vegetatively, creeping outwards from the centre whenever a brief shower of rain permits expansion. As the oldest tissues die off, hollow rings are formed, some of which have been estimated to be between 10,000 and 12,000 years old, which probably makes them the most ancient living things on the planet.

The second type of root adaptation is found in regions where drought is prolonged and the only available water lies deep within the soil, such as that lying beneath the fossilised river beds of the Sahara and central Australia. To exploit these hidden resources, some plants have developed a single tap root of enormous length. The dry river beds of central Australia

are lined with ghost gums, coolabahs and river red gums, all of which extend further below ground than above it, in order to tap the reservoirs deep beneath the surface.

As regards leaf growth, and in contrast to the broad, flat leaves of plants growing in regions of more adequate rainfall, some desert plants are microphyllous, that is, they have very small leaves. In addition, the stomata through which the plant 'breathes' are confined to the leaf margins and these are then rolled under to reduce air movement and maintain humid conditions, thus decreasing evaporation. In some cases, these leaves are so small as to be barely noticeable and, in the most extreme cases, they have been reduced to spines.

The flowering of the desert after rain is a spectacular sight, especially in southwest America. Two species of prickly-pear cactus (**4**,**9**) bear flowers around the rims of their paddle-like branches, as does the related beavertail cactus (**6**), while the deep crimson blooms of the hedgehog cactus (**8**) sprout from between the spines. The flowers of the velvet cactus (**2**) and those of the chainfruit and staghorn chollas (**5**,**7**) are borne at the ends of thin branches. Annuals such as yellow *Eschscholtzia* poppies (**1**) and the perennial penstemon or beard-tongue (**10**) flourish between the cacti.

These spines cannot photosynthesise, that is, they cannot manufacture carbohydrates for building new plant tissues, using energy from the sun and a green pigment in the cells known as chlorophyll, in the same way as normal leaves, so the stem takes over this function. Such is the case with the prickly pear cactus and related species.

The sclerophyllous plants of the arid regions have larger leaves, but these have thick, waxy cuticles, to reduce water loss. The vegetation of the Mediterranean region is dominated by sclerophyllous trees and shrubs, amongst them the carob tree, Kermes and holm oaks, strawberry trees and wild olive trees. Many of these species are evergreen, which allows them to begin

photosynthesis the moment rain starts to fall, rather than having first to produce new leaves, in the manner of deciduous species.

Mediterranean-type vegetation is not confined solely to southern Europe and North Africa, but is also widely distributed between the latitudes of 30° and 40° north and south of the equator. In southern California, for example, the evergreen vegetation of the semi-arid lands is known as chaparral, whilst along the coast of southwest Australia it is called mallee scrub and is dominated by tough spinifex grasses, with inrolled leaves and extensive root systems, growing beneath squat, bushy eucalypts. The latter's predominant leaf colour is grey-green, which helps to deflect the sun's

A shrub very typical of the Mediterranean arid lands is the gum cistus (**11**), with its large white flowers up to ten centimetres in diameter. Its close relative *Halimium ocymoides* (**3**) grows on sandy soils in western and central Iberia.

burning rays. In South Africa, large parts of the Karoo plateaus are covered with blue-green sclerophyllous species and in those parts of Chile, along the coast and in Central Valley, which are subject to a Mediterranean-type climate, similar vegetation can be found, dominated by tree species which grow nowhere else in the world.

There are other types of desert plants which lose their leaves in periods of drought and become dormant. The ocotillo, or coach-whip, of the southwest American deserts is normally a bare and thorny shrub, but, after rain, tiny leaves sprout from the branches. Both the ocotillo and the Australian mulga tree, a kind of acacia, have a very distinctive form. All their branches radiate outwards from a single growing point at ground level and cover a huge surface area. When it rains, water is funnelled down these branches to the point where the plant can make most use of it, the roots. The mulga also has very small leaves and can even drop its branches in severe drought, entering a state of 'suspended animation' to conserve water.

The ability to lie dormant has also been exploited with great success by the so-called 'resurrection plants'. One such species is an Australian liverwort, related to the moss family, which shrivels up and appears to die in times of drought, but springs back to life with the new rains, sometimes decades later. Better known is the crucifer called the Rose of Jericho, which folds all its branches inwards in dry conditions to form a ball, thereby protecting the sensitive reproductive organs of the plant. Its spherical shape allows it to roll with the wind, increasing its chances of finding moister conditions, when it can uncurl and take root once again.

In order to prevent excessive overheating of their tissues through prolonged exposure to the sun, some plants have developed whitish or greyish leaves to

1

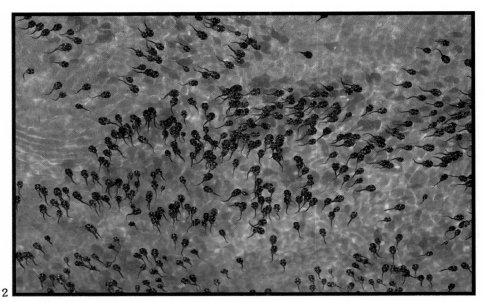

2

During the rainy season shallow pools form and are taken advantage of by creatures needing water to reproduce, such as the bullfrog (**1**). Tadpoles (**2**) are quick to hatch and mature before the water evaporates beneath the relentless sun.

3

4

reflect the sun's rays. The desert holly of the North American arid lands absorbs mineral salts from the ground which it later secretes through the leaves to form a white, powdery coating which serves a similar purpose. These leaves also avoid excessive exposure to the sun by being oriented at an angle of 70° to the vertical so that the noonday sun only strikes their edges. The tamarisk of the Somali Desert also secretes mineral salts, but for a different purpose. The drops of calcium chloride exuded by the leaves are hygroscopic, absorbing water from the atmosphere at night, this moisture later being absorbed by the plant.

In desert regions where evaporation exceeds rainfall, the predominant water movement is upwards through the soil, drawing mineral salts with it. Although the upper layers of soil thus become extremely salty, some forms of vegetation are still able to survive. Saltbushes and bluebushes, which were probably once coastal plants, survive through their ability to eliminate excess salts absorbed through their roots. They do this in such

a way that the soil around the plant becomes so saline that no seeds can germinate, not even those produced by the parent plant. It is only when the rains come and wash away this excess salt, thus ensuring optimal growth conditions for the young plants, that the seeds are able to germinate.

Of all desert plants, perhaps the best adapted to the harsh desert environment are the cacti. These strange plants, with their myriad different growth forms, are confined to the New World. There are about 2,000 species of cactus in the Americas, ranging from the enormous columns of the giant saguaros and organ-pipe cacti to the diminutive, spherical echinocactus species. The giant saguaro is one of more than 140 known species in the Sonoran Desert. Basically it consists of fifteen metres of water-storing tissues, supported internally by rigid, rod-like structures and externally by a thick, waxy skin. The roots are shallow and extend over a large area and, when it rains, those of the largest specimens are capable of absorbing up

Wildfowl are among the first to avail themselves of the wealth of brine shrimps and algae which appear with the first rains that flood the Etosha saltpan in Namibia. Egyptian geese (**3**) and red-billed teal (**4**), both found only in Africa, are unusual among wildfowl in that both sexes have similar plumage. Lesser flamingoes (**8**) also flock to Etosha following the rains.

5

6

The small permanent pools of the North American arid lands are home to desert pupfish (**7,9**). Since the pools lie at some distance from each other these fish have evolved in complete isolation and thus each pool contains a unique species.

7

9

The American xerophytic shrub *Yucca treculeana* (**6**) is renowned for its symbiotic relationship with the yucca moth *Pronuba*. The adults lay their eggs inside the ovaries, at the same time fertilising the flowers, which results in the production of thousands of seeds, some of which serve to feed the developing caterpillars. Lilies (**5**) thrive in arid habitats as their underground bulbs are full of stored food reserves and they can flower with the first rains.

8

to one tonne of water in a single day. The leaves have been reduced to spines and the plant's exterior is pleated longitudinally, not only to increase the surface area available for photosynthesis, but also to allow the stem to expand, concertina-fashion, and thus increase its water storage capacity.

The columnar or spherical forms adopted by many cacti combine this maximum capacity for water storage with a minimum surface area through which this water can later be lost. If this does not appear to be the case above ground, then it is likely that the cactus has a swollen tap root instead, as do some species of prickly pear. This is a particularly true of cacti growing in cold deserts, where the freezing winter temperatures would rupture any water-storing tissues above the surface and thus kill the plant.

In those parts of the world where cacti have never

1

2

3

4

Aloes are characteristic South African desert plants and may have either a single large leaf rosette (**1**) or many smaller shoots each terminating a branch, as in *Aloe dichotoma* (**7**), also known as the kokerboom or quiver tree as its branches provide Bushmen with quivers for their poison arrows. Predatory invertebrates of arid lands include the African baboon spider (**4**) and the European mantis *Empusa pennata* (**6**).

existed, other plants have nevertheless evolved in an almost identical manner, witness particularly the giant euphorbias, or spurges, of Africa and Asia, which are almost indistinguishable from the cacti of the Americas. This process of convergent evolution occurs when completely unrelated groups of plants or animals, usually in different continents, have independently evolved similar life-forms in response to the same type of environmental conditions. The similarities are sometimes astonishing. Like the cacti, African aloes have also developed water-storing tissues in their stems, as have the bromeliads of the caatinga communities of Brazil. Brazil is also home to the water-storing bottle trees, known locally as barrigudos, which swell up to five metres in diameter when full of water. The equivalent African species is the colossal baobab, and there are other similar species in the arid lands of

winds and bitter cold of their present environment that they would be unable to survive in any other.

The best known of these ancient plants, however, is undoubtedly the welwitschia of the Namib Desert. Although a member of the same family, the gymnosperms, the welwitschia is, unlike them, a true desert species and only resembles these conifers in its half-metre flower cones. It is probably an evolutionary relic and has no close relatives in the world today. The bulk of the welwitschia is a large, swollen root, ideal for water storage, which protrudes from the surface of the sand and may be several metres in circumference. From the top of this root grow two leathery, strap-shaped leaves, tattered and frayed by many years of exposure to the elements. The oldest of these plants are thought to have been in existence for over 1,000 years and a more bizarre organism is hard to imagine.

A particularly diverse selection of reptiles occurs in the world's arid lands, ranging from the bizarre but harmless thorny devils (3) and bearded dragons (2) of Australia's heartland to the highly venomous prairie rattler (8) and its close relative the western diamondback rattlesnake (5) of southwest America.

5

6

7

Australia. Because of their slow growth and frugal use of resources, many plants in the arid lands reach a considerable age. The famous grove of Mediterranean cypresses, or tarout trees, of the Tassili Plateau in the heart of the Sahara, for example, contains individual specimens which are thought to be 2,000 to 3,000 years old. They have existed since the times when this part of northern Africa was less arid and, although they still produce viable seeds, these cannot germinate in the present harsh climate. When felled, one ancient specimen of the bristlecone pines from the mountains of America's arid southwest was found to be 4,900 years old. These dead-looking trees have probably reached the end of their evolutionary line, since they are now so well adapted to the scant oxygen, endless

8

Some plants, instead of enduring the long periods of drought which are characteristic of arid lands, choose to flower briefly, set seed and die. The survival strategy of these ephemeral species lies in producing vast quantities of drought-resistant seed, most of which is consumed by insects and small vertebrates, but a small proportion of which escapes and then remains dormant, often for decades, until the next rains. In almost any arid region of the world, the onset of the rains is accompanied by the blooming of its barren landscape. Almost overnight, the monotonous sands and stony plains are covered with brightly coloured flowers which complete their life cycle within a matter of days, sometimes only hours. It is estimated that the woolly plantain, for example, produces some five billion seeds per hectare. Even the saguaro cactus, although not an ephemeral species, may produce up to forty million seeds in its lifetime, of which only three or four will survive both predators and the rigours of the climate to develop into mature plants.

Since a brief shower would probably provide insufficient water for the adult plants to reproduce successfully, the seeds of these ephemeral species are coated with a variety of germination inhibitors, which ensures that they cannot start to grow until the rainfall has been heavy enough to wash these inhibiting chemicals away. The plants then develop rapidly and their flowers are brilliantly coloured and heavily scented to attract pollinating insects as quickly as possible. In addition, the plants often adopt a creeping habit so as to extract the maximum amount of moisture from the quickly drying sands. Other than in their seeds, however, these ephemeral species display none of the xeromorphic adaptations of the perennial desert plants.

No living creature can survive without water. It is essential for all life processes and constitutes the major proportion of the tissues of any animal or plant. Despite this, many desert animals never drink at all, but obtain all the moisture they require from their food. Herbivores, such as the desert cottontail of the Sonoran Desert or the Australian hairy-nosed wombat, manage to obtain sufficient water for their survival from the vegetation on which they feed. Similarly, the carnivorous creatures of the desert obtain the water necessary for their survival by consuming their prey, whose bodies consist of at least two thirds water. These predators include the coyotes and bobcats of the North American deserts, the diminutive marsupial kowaris and mulgaras of Australia and the now extremely rare Asiatic cheetah of the Turkestan Desert. They also include birds of

The Australian frilled lizard (1) is one of the continent's most spectacular reptiles. Its long, slender body measures up to ninety centimetres, although about two-thirds of this length is tail, and it is mainly arboreal, feeding on small vertebrates and insects. When threatened, however, the frilled lizard will take to the ground and run off on its hind legs, head tilted back and front limbs clasped to its chest like a miniature tyrannosaur. On reaching what it regards as a safe refuge, the lizard turns to face its enemy and opens its mouth, thus extending a large expanse of skin supported by cartilaginous rods in the manner of an umbrella, which serves to shock the pursuer.

Range of the Frilled Lizard

The savanna monitor lizard (2) is widespread throughout Africa's deserts, as is the baobab (10), one of the trees best-adapted to the continent's arid conditions. Little water can be lost through its sparse, small, leathery leaves, and its barrel-like trunk is able to store a large volume of fluid for gradual use during the drought. The handsome *Phlomis lychnitis* (4), a member of the mint family, is found in dry rocky terrain in southern and central Iberia.

prey, such as the wedge-tailed eagle and grey goshawk of Australia, the lanner falcon of the Sahara and the world's widest-ranging diurnal land bird: the peregrine falcon.

In areas where rainfall is particularly scarce, all possible sources of water must be exploited. Thus the opportunistic darkling beetles of the Namib Desert take full advantage of the mists, which regularly sweep in from the South Atlantic, by climbing to the crests of the sea-facing dunes well before dawn and doing headstands in the sand. This is not as ludicrous as it may seem; all the bristles and grooves on the tough underside of the beetle lead eventually to the head, so, by presenting the maximum body area to the incoming

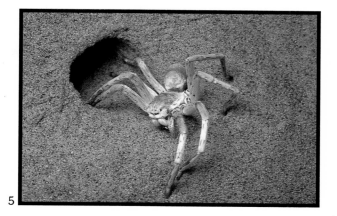

5

Scorpions (**9**), of which there are some 650 species in the world, are particularly abundant in desert regions. In most species the young are born live and the level of parental care is surprisingly high, the juvenile scorpions travelling around on their mother's back until after their first moult. In the Namib Desert the white lady spider (**5**) is a particularly voracious predator, emerging from its burrow at dusk to hunt. Even creatures as large as geckos are numbered among its victims. The arid lands of Eastern Australia are home to the common bearded dragon (**7**). (**6**) The 'prehistoric' features of an iguana.

6

7

8

9

10

The Australian shingleback lizard (**8**) has a short, stumpy tail which is very similar in outline to its head, thus confusing aerial predators. The adults can measure more than thirty centimetres from head to tail and the newborn are, incredibly, well over half this length. The desert tortoise (**3**) of the Sonoran Desert and other arid regions of southwest America is able to survive long periods without water, obtaining all of its moisture from the vegetation it consumes.

mist, it can trap the moisture and funnel it down towards the mouth. For obvious reasons, these beetles, are sometimes also known as head-standers. Similarly, the thorny devil lizard of Australia's arid lands is thickly covered with ridges and spines, which, despite their intimidating appearance, serve only to direct rain or condensed dew towards the lizard's mouth.

Other animals have changed physiologically. Instead of wasting precious water passing urine, most desert lizards produce dry crystals of uric acid. Camel dung is so dry that it can be used as fuel almost immediately! Over the course of many thousands of years the kidneys of the kangaroo rat of the Sonoran Desert, where the average annual rainfall is less than 250 millimetres, have become extremely efficient, allowing this small mammal to excrete a paste rather than pass liquid urine. The kangaroo rat has two other water-saving devices: firstly, it has no sweat glands and secondly, its nasal passages are maintained at a temperature a few degrees lower than the rest of its body so that, when the animal exhales, any moisture in its breath condenses in these passages and is thus retained within the body.

One way of avoiding the extreme surface temperatures experienced in many arid regions of the world is to adopt a nocturnal existence, hiding from the penetrating rays of the sun during the day in underground burrows. In sand dunes, for example, the surface temperature may be a scorching 80°C, yet, a metre below ground level, it is a comfortable 23°C. Understandably, a great many creatures, including kit foxes, kangaroo rats, gerbils and jerboas, take refuge in this sub-surface realm, retreating there well before dawn. An added advantage is that the humidity below the surface may be six times as great as on it, due largely to the animal's own respiration. Once formed, this humid atmosphere reduces any further loss of body fluids.

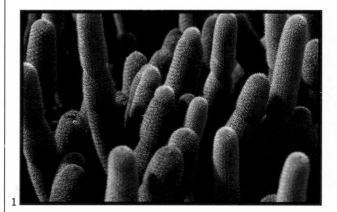

1

2

Desert animals have evolved many other techniques to avoid overheating. Some lizards stand upright, to raise their delicate underbellies further away from the intense heat radiated from the sand's surface. For the same reason, other lizards run about on tiptoe and yet others raise their legs alternately in diagonal pairs, front left with back right and vice versa, looking for all the world like four-legged tightrope walkers. The spiny-tailed lizards have undergone certain physiological changes and can now survive body temperatures which may reach an incredible 47°C. The Kalahari ground squirrel creates its own shade by using its bushy tail as a parasol, and this enables it to continue feeding throughout the day. Yet other creatures, such as the dancing spider of the Namib Desert and some Saharan species of ant, are covered in fine white hairs, which reflect the sun's rays.

Large ears, another feature common to desert animals all over the world, are characteristic of such diverse creatures as the kit foxes and antelope jack

3

4

The massive baobab (**3**) grows throughout southern and tropical Africa, the largest specimens reaching heights of over twenty metres, with boles up to forty-three metres in circumference. The stems of the cactus *Brachycereus* (**1**) are covered with a dense coating of fine spines which are, in fact, reduced leaves. Some forty species of yucca grow in the southern United States, the Mojave yucca (**4**) being confined to the desert of the same name. Narrow-leaved cistus (**2**) is an aromatic shrub typical of the western Mediterranean maquis.

rabbits of the southern United States and northwest Mexico, the African fennec and bat-eared foxes, the long-eared hedgehogs and jerboas of the Kara Kum and the Sahara Deserts, the Australian rabbit-eared bandicoot and, of course, the African elephant. The antelope jackrabbit, more accurately a hare than a rabbit, is perhaps the best example, with ears up to twenty centimetres long – one quarter of its body length. A network of tiny blood vessels runs very close to the ears' surface so that, when the animal faces north and raises its ears into the breeze, the circulating blood is cooled and the temperature of the whole body is lowered. When the jackrabbit wishes to conserve heat, at dusk or overnight, it simply folds its ears back over its body.

Kangaroos, although they too have large ears, are unable to lose sufficient heat in this way and so they have an additional network of these blood vessels on the inside of their forearms. In very hot conditions the kangaroo licks its fur at this point, working up a lather with saliva; this increases evaporation from the forearms, thus cooling the blood and lowering the animal's body temperature.

A period of dormancy, during which an animal's body temperature and metabolic rate are greatly reduced, is employed by certain species to survive severe droughts without needing food or water. The hind limbs of the spadefoot toad in the Arizona Desert have been modified into flat, shovel-like appendages which help the toad to burrow backwards into the sand, at the onset of drought. Once buried in this cool, moist burrow, up to half a metre below the surface, the outer layer of the toad's skin hardens to form an almost watertight cocoon, with just two small holes at the nostrils to allow the toad to breathe, and here it remains until the drought is over. In Australia, the catholic and burrowing frogs react in the same way to prolonged drought.

American desert reptiles include the Gila monster (**5**), the banded gecko (**10**), the collared lizard (**11,15**), the regal horned lizard (**9**) and the eastern hog-nosed snake (**8**) as well as more than thirty species of rattlesnake, including the Mojave rattler (**6**) and the banded rock rattler (**12**). The horned viper (**13**), the burrowing skink (**14**) and the similar *Mabuya maculilabris* (**7**) are all to be found in Africa.

A period of dormancy is also employed by animals in the deserts of the northern hemisphere. In their case, however, it is not in response to drought conditions but to the bitter cold of the winters, when food is hard to come by. The most extreme example of desert hibernation must surely be the poor-will, a member of the nightjar family, which lives in the deserts of the southwestern United States and northern Mexico. Most birds are unable to slow their metabolic rate sufficiently to survive the severe winters, but the poor-will is able to hibernate, barely breathing, for up to three months of the year and thus avoid the intense cold. No-one is quite sure how it manages to lower its body temperature from 42°C to a mere 18°C, but, even more astounding, it can do so at a moment's notice, without requiring, as most mammals do, a period to build up fat reserves. It is, however, a species of desert tortoise inhabiting the Kara Kum Desert which has taken this way of life to extremes, lying dormant throughout both the cold winter and the hot summer and emerging only for a few brief weeks in April and May to reproduce and then to build up its fat reserves for the next period of hibernation.

Another common strategy in this uncertain environment, employed by creatures as diverse as the Gila monster of the Sonoran Desert, the Australian rat kangaroo and small mammals, such as the fat-tailed jerboa of central Asia, is to build up fat reserves during times of plenty and store them in their tails. In the case of the Gila monster, one of only two venomous lizards in the world, this fat is then slowly metabolised during hibernation.

1

2

3

Moving over the scorching desert sands presents problems, but some ingenious behavioural adaptations have overcome these. The sidewinding adder of the Namib Desert has evolved a method of locomotion involving sideways flicks of the body so that only two or three short sections are in contact with the ground at any one time. Independently, an unrelated snake from the deserts of the southwestern United States, the sidewinder rattlesnake, has evolved a very similar technique in response to almost identical conditions and it can move at speeds of up to four kilometres per hour across the sands. This rattlesnake, about forty-five centimetres long and well-camouflaged, hunts at dusk or during the night, actively locating its prey by means of heat-sensing pits in the side of the head and a tongue which is highly receptive to even the faintest of smells. By contrast, the sidewinding adder prefers to bury itself in the surface layers of the sand and ambush unsuspecting prey. Both species have fast-working

4

5

6

Even at the height of the dry season the Etosha Pan is not without life. Desert raptors such as martial eagles (**1**) and lanner falcons (**6**) are regular visitors and gemsbok (**3**) and ostriches (**2**) frequently cross the salt-laden plains. Africa's dwarf mongoose (**4**) is renowned for its snake-killing abilities. The creosote bush (**5**) forms part of the typical desert scrub of southwest America.

venom; they also swallow their victims whole and then digest them later, at leisure.

The bounding movement employed by so many desert mammals is also designed to minimise contact with the hot sands. In addition, it has been shown to be energy-saving, which can be a critical factor where food and water are widely separated, as is often the case. The fawn-hopping mouse and kangaroo of Australia are good examples of creatures which move in this way, as are the gerbils and jerboas of the Sahara and their equivalent in the Sonoran Desert, the kangaroo rat. The success of this technique is demonstrated by the existence of no less than sixteen known species of jerboa in central Asia, nine species of hopping mice in Australia's arid interior and five species of kangaroo rat in Baja California. Movement across such an unstable surface as the loose, shifting sands of many deserts presents different problems, to which solutions have also been evolved. Many creatures have increased their foot area so as to spread their weight more evenly and widely and thus avoid sinking into the soft sands. The North African addax antelope, for example, has developed splayed feet and some species of gecko, such as the snowshoe gecko of the Gobi Desert, have evolved webbed feet, much like those of frogs and toads. The brush-toed jerboa of central Asia and the African sand-cat have very hairy feet, which increases their surface area and thus the animal's ability to traverse loose sand in an efficient manner. The roadrunner, a half-metre-long member of the cuckoo family, has feet whose toes are spread like a cross to give good traction on the loose surface. This comical bird of the southern United States and the Mexican cactus deserts has a tail like an exclamation mark, which it uses for steering and braking. It can reach

1

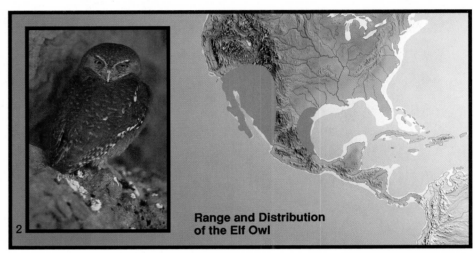

The yucca known as the Joshua tree (**7**), which may attain some fifteen metres in height, is a common feature of the American southwest, along with giant saguaros (**5**) and organ-pipe cacti (**6**), all of which provide valuable feeding and nesting sites for the desert avifauna.

2

Range and Distribution of the Elf Owl

3

4

5

6

7

The summer tanager (**8**) is widely distributed throughout the dry forests of North America, including the peripheries of the Sonoran Desert. The American nightjar, known as the poorwill (**9**) for its monotonous two-tone call, is unusual among birds in that it is known to enter a true state of hibernation during the winter. Red-tailed hawks (**4**) and elf owls (**2**) both make use of the saguaro for breeding purposes, the latter occupying holes previously excavated by the Gila woodpecker. In the same region, Gambel's quail (**1**) rears very large numbers of offspring when the rains arrive on schedule, but in years of drought they do not reproduce.

8

9

speeds of up to twenty-five kilometres per hour in short bursts and, although it has wings, prefers not to fly.

Other species have given up trying to keep their feet on the ever-shifting surface and have opted instead to 'swim' through the sand using S-shaped flexures of the body in the manner of a snake. Lizards such as skinks and shinglebacks have become more streamlined over the years and their heads more wedge-shaped. These lizards have also evolved transparent eyelids to protect their eyes from damage by abrasive sand grains, whilst still allowing them to see where they are going.

As protruding limbs are unnecessary for, and may even be a positive hindrance to, this type of locomotion, they have been lost altogether in several species of lizard. In Australia, for example, skinks of the genus *Lerista* show both stages in this evolutionary saga; *Lerista bipes* retains vestigial hind legs, the forelimbs having been lost long ago, whilst *Lerista apoda* has lost even these in the process of becoming better adapted to its environment. The Australian flap-footed lizards have also found legs to be superfluous and can

10

The griffon vulture (**3**) is a typical scavenger of arid lands around the Mediterranean. Each vulture soaring across the plains in search of food also keeps an eye on its neighbours so that when one bird descends to an animal carcass the rest follow. Unlike most falcons, the lanner falcon (**10**) adopts a gliding flight to conserve energy, feeding mainly on grounddwelling vertebrates as it lacks the weight to knock a bird from the sky.

1

2

The Gila woodpecker of the Sonoran Desert is one of about 210 species of the family *Picidae*. The male (**7**) sports the black and white plumage characteristic of most of these birds, but the female (**2**) is, unusually, predominantly brown.

3

4

now only be distinguished from snakes by their fleshy, as opposed to forked, tongues, and by the fact that, like all lizards, they are capable of shedding their tails to confuse and thus evade would-be predators. Finally, as they are not venomous, they have to resort to immobilising their prey by biting its legs off!

The masters of the subterranean desert world are undoubtedly the golden mole of the Namib Desert and the marsupial mole of Australia. Although not even closely related, these two creatures show remarkable similarities both physically and in their life-styles – a true example of convergent evolution. Both animals have broad, wedge-shaped noses to help them burrow through the abrasive sand with the difference that, as the marsupial mole lives in rockier habitats, its nose is additionally protected by a horny shield. Both animals also have golden coats, whose fur completely covers their eyes, these being useless in their subterranean world, and their ears are simply small holes. In the golden mole, the limb-bones are buried deep in its flanks, leaving only the feet projecting, but the marsupial mole has evolved enormous claws on its forefeet to facilitate burrowing in the terrain in which it lives.

Reptiles are possibly the animals best suited to an arid environment, as can be seen from the large numbers of different species which inhabit the various desert regions of the world. There are no less than eighteen species of rattlesnake in Baja California, most

5

Bee-eaters are insectivorous birds that especially favour dragonflies, beetles, butterflies and, of course, wasps and bees. They usually trap insects on the wing, often by perching on a convenient branch – or telegraph wire, if such a vantage point is available – from which they launch themselves after passing insects. The carmine bee-eater (**1**) of the Sudan, however, adopts a rather different hunting method, hitching a ride on the back of a bustard (**5**) and making aerial forays to pick off insects disturbed by the progress of the larger bird.

of which is arid terrain, and there are an astonishing forty-five snake species in Russian Turkistan. Of a total of 500 lizard species in Australia, about 200 live in the country's arid interior, and one square kilometre of sand-ridge may support as many as forty different species. Amongst these Australian species are some of the world's largest lizards, including the predatory sand monitors and perenties, which grow to nearly two metres in length.

Reptiles require very little water and do not generally need to drink. Their great disadvantage is that they have no internal mechanism for regulating body temperature and so they are dependent on the efficient use of sun and shade to achieve both the low body

temperatures ideal for resting and sleeping, when metabolic rate slows and little energy is used, and the higher temperatures required to move quickly and hunt efficiently. So, early in the morning, after a bitterly cold desert night, lizards and snakes will emerge from their shelter and lie broadside-on to the rays of the sun, in order to absorb as much heat as possible and thus raise their body temperature. Later in the day, however, any reptile still on the surface will carefully orientate its body so that it is facing the sun, thus reducing the area struck directly by its rays.

By contrast, the great disadvantage of most desert birds is that they are unable to dig tunnels to escape the heat of the sun. The diminutive burrowing owl has

The rollers of the Old World resemble brightly-coloured crows and are named for their habit of turning somersaults during the courtship display. Both lilac-breasted rollers (6) and broad-billed rollers (8) are found in eastern Africa, where they feed on small, ground-dwelling vertebrates and insects.

6

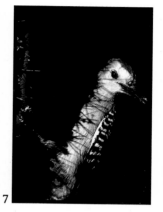

7

8

The ostrich (4) of Africa is the world's largest bird, reaching a height of almost two-and-a-half metres, and the Australian emu (3) is only marginally smaller. Although superficially similar, these two flightless birds of the open plains belong to different orders. In the Kalahari the South African Bushmen collect ostrich eggs and fill the emptied shells with water. These are then buried at strategic points across the desert, and can make the difference between life and death to a Bushman hunting party.

overcome this problem by sharing the tunnels of other desert creatures, often living in colonies in the burrows of rodents and even lizards. The saxoul jay of the Kara Kum Desert, a member of the crow family sometimes known as Pander's chough-thrush, also makes use of rodent burrows, particularly in the nesting season. Another disadvantage of desert birds is that they lose a lot of moisture in their efforts to keep cool, which the majority do by opening their beaks and vibrating their throats. An alternative method of lowering body temperature involving less moisture loss is employed by the ground-dwelling Australian gibber bird, which holds its wings slightly away from its body as it runs, allowing the increased air movement to cool its blood.

In an environment that offers so little opportunity for concealment as the desert sands, one might think that predators would have few problems in hunting down their prey. However, the hunted have evolved many varied and ingenious methods by which to avoid their hunters. For those creatures obliged to remain on the surface during the day, perhaps because they need to feed continuously, or cannot dig burrows, camouflage is extremely important and the animals of the world's arid lands display some of the best examples of cryptic

coloration in the animal kingdom. In Australia, for example, the gibber dragons are rust red, as are the plains where they live; the Lake Eyre dragons, on the other hand, are almost white, the predominant colour of the saltpans of their low-lying environment.

Other animals have evolved more active means of escape from their predators. The many species of long-legged desert rodent, and their marsupial

1

Deep in the Namib Desert grows one of the most extraordinary plants in the world: the conifer welwitschia (**1**). Like many arid-land species, it possesses a long, thick taproot to extract subterranean moisture and store it. The American burrowing owl (**2**) is quite capable of excavating its own nest holes, but prefers to occupy those of other creatures, such as prairie dogs in the north and viscachas further south, rearing its chicks on the newborn rodents.

2

3

4

True desert birds are generally clothed in drab plumage so as to escape detection in a land where few hiding places are available. Many, including the African kori bustard (**3,11**), the roadrunner (**10**) of the arid American southwest and the Australian mallee fowl (**8**), lead a largely terrestrial existence, while saker falcons (**9**) are among the most skilled aerial predators of the desert. (**5**) Young Namaque sandgrouse quench their thirst by sucking the breast feathers of their father, and immature stone curlews (**6**) shelter from the scorching sun between their parents' legs (**7**).

equivalents in Australia, are able to change direction in mid-leap to throw their attackers off course. The antelope jack rabbit employs another, different tactic; as it zig-zags away from a pursuer, it contracts the muscles immediately beneath the skin so as to lift the hair on its flanks and expose its white under-fur. This technique, known as 'flashing', is thought to confuse the predator and is certainly a conscious tactic, rather than an automatic reaction on the part of the jack rabbit, which has been shown to be able to control its use.

Predators, in turn, have developed various strategies to outwit their victims. Most obviously, birds of prey have evolved incredibly sharp eyesight and are able to spot their prey from a position high in the sky and then swoop down very suddenly to secure it. Many of the ground-dwelling hunters are very fleet of foot and possess lightning reflexes. But the most successful predators are perhaps those which choose to conserve their energy, preferring to wait for their victims to come to them. The horned viper of North Africa and Arabia is one of the most venomous of all desert snakes. It is stout and sluggish and thus unable actively to hunt down its prospective victims; instead it half buries itself in the sand and waits, springing suddenly from its hiding place to sink its fangs into any unsuspecting lizard or rodent which might pass by. The sand boa of the Kara Kum Desert is particularly well adapted to this mode of life, having periscopic eyes

The head of the Egyptian vulture (**12**) is covered with long, cream-coloured feathers which are difficult to clean, unlike the bald pates of most vultures, so these small black and white scavengers usually hover on the sidelines of the feast at an animal carcass, waiting for scraps to be tossed aside. Huge flocks of sandgrouse (**4**), disturbed by grazing animals, perform elaborate aerial acrobatics that recall the precise coordination of shoals of fish.

The African aardwolf (**2,3**) favours sandy regions such as the Kalahari. It is related to the hyena, but is possessed of a much weaker dentition – it feeds predominantly on termites and has no need of bone-cracking teeth.

which can scan the horizon for possible victims even when the snake is completely buried under the sand.

Other masters of the ambush technique are the ant-lions, the adults of which are harmless winged insects, resembling lacewings. The larvae, however, are armed with vicious, calliper-like jaws and have a voracious appetite. They excavate a steep-sided, conical pit in the sand, conceal themselves at the bottom and await their victim. Any creatures, in particular ants, which unwittingly slip over the edge of the pit will, in their efforts to climb out, send sand grains rolling down the pit's steep sides, thus alerting the ant-lion larva below.

The ultimate function of all living things is to reproduce themselves, thereby ensuring the continued existence of their species. In arid areas, the opportunities for procreation may not only be brief but also few and far between. Many years may pass between the rains which bring the optimal conditions for reproduction and it is therefore essential that species have some means of surviving these long periods of drought.

A rapid response to favourable conditions is also useful to take full advantage of brief periods of rain. The spadefoot toads, roused from their burrows by the sound of raindrops on the ground above, are a good example of this. Within a few hours of emerging from these burrows, they have mated and laid their eggs in the newly-formed pools. Within twenty-four hours these eggs have hatched into tadpoles. The next stage depends on the food supply in the pool and on the

length of the wet period. At first, the tadpoles feed on the algae that will have developed from some of the microscopic spores which are always blowing around the desert in search of water. If fairy shrimps, whose drought-resistant eggs are blown around the desert in much the same way, also develop, some of the tadpoles will grow huge heads and wide mouths to feed on them. The spadefoot toad is thus preparing for two eventualities. If there is no further rain and the pools begin to dry out, the carnivorous tadpoles will develop rapidly, eating their smaller, herbivorous brothers and competing to establish themselves in the deepest parts of the shrinking pools. Only the largest and most

The impressive century plant (**1**) of the North American arid lands is a feature of the landscape that is home to the diminutive kit fox (**4**). The huge ears of this nocturnal hunter act as heat exchangers to lower its body temperature.

The honey possum (**6**) of the southwest Australian plains has been described as the 'marsupial hummingbird' because of its nectar-sipping habits. It is less than eight centimetres long and possesses a slender, prehensile tail with which it can maintain its grip while feeding.

6

7

5

Although usually considered to be creatures of the African savanna grasslands, cheetahs (**5**) and lions (**8**) are equally at home in more arid regions such as the Etosha Pan in Namibia. Meerkats (**7**), known in Spanish as 'little men of the desert', are also denizens of southwest Africa. These small carnivores, close relatives of the mongoose, are often seen standing on their hind legs, using their tails for support, on sentry duty. They are unusual among carnivores in having an extremely well-developed social organisation, with cooperation in hunting, care of the young and the defence of the burrows against enemies.

8

aggressive of them will survive to become toads. On the other hand, if further rain falls, stirring up the waters of the pool so that they become murky, the carnivorous tadpoles will be at a disadvantage, since hunting is more difficult if you cannot see your prey. The herbivorous tadpoles, however, will continue to feed on the algae and will gradually develop into tiny toads; although smaller, fewer of these will die, because there is an abundant supply of food and the threat of imminent drought has been removed.

For the larger mammals, life would seem to be insupportable in the harsh desert environment. Some have nevertheless adapted to the conditions there with remarkable facility and are not now found anywhere else. The Indian wild ass, or ghor-khar, for example, is found only in the Little Rann of Kutch Desert in western India. These wild asses have white undersides to reflect the heat from the scorching sands, and, in contrast to many denizens of the desert, they do not need to seek shade during the heat of the day, tolerating temperatures of up to 44°C. They also rarely need to drink, quenching their thirst by eating the succulent,

1

2

3

4

shrubby seablite which grows plentifully in this saline desert. Further north, the related Mongolian wild ass, or kulan, can be found. The Badkhyz Reserve, totalling 75,000 hectares, was set up specifically to protect the last 150 desert-dwelling kulan in Soviet Turkistan; today there are about 1,000 individuals, with a further 500 in Afghanistan, which, despite the fact that even in summer kulan can survive for up to two or three days without drinking, is close to the upper limit that the available water supply can support.

Other wild asses have not been so fortunate: the North African wild ass, which once roamed the Atlas

Mountains, has been extinct for some time, whilst the Nubian and Somali wild asses are now confined to the hills around the Red Sea. In fact, the Nubian wild ass may also be extinct, through crossbreeding with the local domestic ass, but about 3,000 of the Somali wild ass do still exist in Ethopia and to a lesser extent in Somalia, where they are strictly protected. They have an exceptional resistance to thirst, needing to drink only irregularly, an ability which also allows them to avoid possible predators at waterholes.

Of all the large animals, the addax, a species of antelope, is perhaps the best adapted to desert

The desert race of the wide-ranging leopard (**1**) is lighter in colour than its forest counterpart and is also much smaller, weighing around sixty kilogrammes as opposed to over ninety kilogrammes. The caracal (**5**), also known as the desert lynx, is largely confined to arid regions, ranging from Africa to southwest Asia; it has a much longer tail than other lynxes, and larger ears. The related bobcat, or bay lynx (**8**), is often seen on the fringes of North American urban areas, where it feeds on rats and mice, though it is also capable of bringing down an adult deer. Other feline predators of the world's arid lands include the slender jaguarundi (**4**) of the American southwest and the diminutive, arboreal African wildcat (**7**).

5

6

The desert cottontail (**3**) relies more on avoiding the worst of the heat than on any specific physical adaptations to its arid environment in southwest America. The vivid crimson bracts of *Castilleja coccinea* (**6**), a member of the snapdragon family characteristic of America's arid lands, give this plant its common name of Indian paintbrush.

7

8

conditions. This animal can even survive in the Saharan ergs, those wildernesses of drifting sand and searing daytime temperatures. Some maintain that the addax never drinks at all, whilst other, more conservative, sources state merely that they can survive for up to a year without drinking, obtaining all the water necessary for their survival from the vegetation they consume. Their hooves have spread and become flattened to facilitate movement across the unstable sands and their eyelashes are long and thick to protect the eyes from both wind blown sand and the glare of the sun. In 1974 there were an estimated 5,000 addax in Chad, but war

and poaching have seriously depleted their numbers. Today they are considered to be an endangered species, surviving only in the Aïr Mountains and the Ténéré Desert in northern Niger.

The Arabian oryx, the smallest of the four antelopes of the genus *Oryx*, is also a true desert creature; it too can survive for long periods without access to water and it uses its thin, sharp-pointed horns, which may be up to seventy-five centimetres long, and its hooves to excavate holes beneath shrubs and at the bases of sand dunes, in which it shelters from the sun. Its coat is so white that in Arabic it is known as *maha*, the crystal.

Although they originated in Arabia, the majority of one-humped camels in the world today live in Africa. There are no wild Arabian camels left in existence, all one-humped camels being of a special domesticated race known as the dromedary (**2**), valued for its powers of endurance in desert conditions.

The last seven wild representatives of this magnificent beast were shot in Oman in 1972, however, due to the foresight of various conservation agencies, a small, captive breeding population had already been established in Phoenix Zoo, Arizona. The breeding programme was so successful that by 1978 there were 120 animals in captivity. The offspring of this small group, known as the 'world herd', have since been successfully released in the Shaumari Reserve at Azraq, Jordan, as part of a programme intended to replace the native fauna of that country, much of which has disappeared since the Second World War. Some have also been released in Oman and Saudi Arabia.

Fifty years ago, five endemic ungulates thrived in the deserts and rocky mountains of Saudi Arabia; the Arabian oryx, mountain gazelle and Dorcas gazelle are all extinct today, and the Nubian ibex and goitred gazelle are restricted to a few remote areas in the mountains. Their decline has largely been caused by the development of the desert following the discovery of oil beneath the sands, and also by the increased mobility and more sophisticated weaponry of trophy hunters. Today, the Nubian ibex is confined mainly to the Sinai Peninsula, although some were recently rediscovered in Egypt's Eastern Desert after an absence

1

2

3

The South African springbok (**1,2,3**) gets its name from its extraordinary response to danger. If the animal is alarmed it drops its head and arches its back, springing stiff-legged up to two metres into the air, while at the same time a dorsal pouch lined with long white hairs is everted, displaying a prominent crest. During the nineteenth century vast migrations numbering millions of springbok occurred regularly, but, in a bid to reduce competition with domestic herbivores, these elegant antelopes were slaughtered wholesale and only a fraction of their former numbers remain today.

of more than sixty years. Another race of goitred gazelle, known locally as *djeiran*, survives in the deserts of central Asia. Here, where temperatures exceeding 50°C have been recorded, they have shown a considerable degree of adaptability to their desert home and, although they were at one time almost extinct, there are now more than 5,000 in the vicinity of Badkhyz.

Perhaps the best known of the world's desert mammals is the camel. Of the two species, the single-humped Arabian camels are now extinct in the wild,

although they were probably the ancestors of the present domestic dromedary, but the two-humped, or Bactrian, camels still survive in Mongolia's Gobi Desert. Camels first appeared in North America, about forty million years ago, and were much smaller animals at this stage. About two-and-a-half million years ago they crossed into the arid regions of Asia and Africa via the land bridge that then spanned the Bering Straits. Here they developed their characteristic splayed feet, for ease of movement across the desert sands, and hump for fat-storage.

Although they became extinct in their place of origin, camels flourished in Asia, continuing to adapt to the conditions in these expanding arid regions. Their woolly coat is a good defence against the heat, slowing the rate of evaporation through sweating, thus cooling the body more efficiently, and also providing good insulation against the cold both at night and throughout the long winters. Long legs raise the camel's body far enough above the sands to protect its vital organs from the intense heat radiating up from the surface; they also account for the camel's loping, energy-efficient gait. According to popular myth, the camel stores water in its hump; although this is not the case, as the hump's fat reserves are slowly oxidised to provide the camel with energy, they do also yield water which can then be used in its metabolic processes. In extreme drought, the camel is able, to a limited extent, to extract water from its blood plasma, although, if fresh water supplies do not quickly become available, the blood becomes too thick to circulate and the animal dies. After a long period without access to water, a camel can drink up to 180 litres in one go.

Other large animals have made behavioural, rather than physical, changes. The elephants which live in the Namib Desert, for example, cannot be distinguished

4

The boojum tree (**6**) of southwest North America, with its solid, cylindrical trunk, short, spiny branches and small, leathery leaves, can withstand long periods without water. Giant spurges and stapeliads, bearing a striking resemblance to New World cacti, dominate the arid South African landscape (**5**). Comparable environmental conditions have given rise to similar life-forms, each having succulent stems and leaves reduced to spines.

6

The African lion (**4**) is not a true desert creature, but is nevertheless widely distributed across the semi-arid lands of this continent. It is only in the last century that the lion has been gradually driven from its natural wooded grassland habitat into less suitable terrain by poaching and increased human activity on the plains. The young males have a full mane, generally paler and less heavy in those which occupy drier lands, by the time they are five years old. Similarly, the Asian elephant (**7**) is not considered to be a desert creature, being more usually associated with tropical forests. In Sri Lanka, however, the destruction of most of its natural habitat has led to the increased use of more arid regions, where sand-bathing becomes necessary to protect its skin from excessive insolation.

5

physically from those which live in Africa's tropical jungles; despite having lived in the desert for centuries, they have not evolved into a separate subspecies, but have instead adopted a life-style more suited to the arid conditions. The number of true 'desert elephants' today has dropped to a mere seventy individuals, largely as a result of poaching, but they have perhaps fared better in the Namib Desert than have their counterparts elsewhere in Africa, since the desert is hostile country for man.

It was once thought that the Namib elephants were migratory, entering the region only after the rains, when vegetation is lush and food is available in abundance. Researchers have since proved, however, that these animals live in the desert permanently. Since an adult bull consumes about 100 kilograms of vegetation daily, the elephants confine themselves

7

mainly to the sand rivers that snake across the Namib Desert, where plant life of some description is always present. Grasses form the bulk of their diet during the infrequent wet spells, but for the most part mopane acacias and commiphoras are their main sources of food.

In contrast to other elephants, those in the Namib Desert only need to drink once every four days or so, which allows them to forage over much greater distances. They dig holes in the parched river beds in order to reach the underlying water; these waterholes are subsequently used by a great variety of other animals, which would not be able to survive without them. In the heat of the sun, the elephants flap their ears gently, which can cool their blood by as much as 6°C, and they may even put their trunks in their mouths, to extract water from the stomach, which they then spray behind their ears to increase this cooling effect through evaporation. Dust-bathing is also a common phenomenon and the coating of reflective particles may help to protect the otherwise vulnerable skin of the elephant from the sun.

The survival of this small herd of desert elephants is thought to be based on their intimate knowledge of their surroundings and, for this reason, the young animals stay with their parents for rather longer than is usual, in order to learn exactly where the waterholes and feeding areas are. It is this same intimate knowledge which makes the herd irreplaceable. Should these particular elephants die out, any attempt to reintroduce the species into the Namib Desert would necessarily fail through the newcomers' lack of specialised knowledge.

True deserts and arid marginal lands support unusually complex ecosystems, with a wealth of animal and plant life which has adapted to the harsh environment over millennia. But arid lands are also home to about 700 million people, and the damage

Large herbivores which are frequently encountered in the southwest African arid lands are the handsome, long-horned gemsbok (**2**) and small herds of desert elephants (**3**). A few dromedaries (**9**) are still to be found in Saudi Arabia, while the predominant large grazing creature of the Australian interior is the red kangaroo (**4**).

they have done through the grazing of domestic animals, through the collection of firewood and through poor agricultural practice has led to the erosion of what little topsoil had accumulated, so that one fifth of the earth's land surface, a staggering thirty million square kilometres, has now undergone, or is in direct danger of, desertification. This is the process by which stable environments are converted into barren wastelands – false deserts, in which no living organism can survive.

More than 100 countries of the world are afflicted by this problem; on a worldwide scale, over forty hectares of marginal land is being destroyed by the process of desertification every minute. The Sahara alone, the world's greatest desert, has invaded more than a million square kilometres over the past fifty years. In the short length of time that man has been on earth, he has already started to destroy the unique desert ecosystem which has taken millennia to evolve.

Two of the more bizarre small mammals of the world's arid lands include the Ethiopian hedgehog (**1**) and Merriam's kangaroo rat (**5**) of southwest America.

6

7

Like all cacti, the cottontop (**6**) and the barrel cactus (**7**) of the American southwest both have leaves reduced to spines and a green stem for photosynthesis. Indian jackals (**8**) are primarily scavengers, although when hunting in packs they are able to separate young herbivores from their mothers.

8

9

· *CHAPTER 3* ·

POLAR REGIONS

The shining icecaps of our world – Antarctica and the Arctic – are among the least known of all the earth's regions. Permanently frozen by bitter, dark winters, they are hostile environments to all but the hardiest men. However, during the summer, when the ice sheets retreat, their shores support the world's largest colonies of seabirds and seals, which congregate here to feed and rear their young. The cold, pure waters of these polar seas teem with life ranging from the microscopic krill to the largest mammal ever to have lived on this planet – the blue whale. As yet, these lands of the midnight sun remain relatively free from the influence of man – the last wilderness areas of our increasingly polluted and devastated earth.

Left: magnificent king penguins in Antarctica, (top) polar bears in the Arctic, and (above) the lesser rorqual or minke whale, which frequents the seas of both regions.

The poles, representing the conjunctions of all the lines of longitude that circumscribe the world, correspond to the ends of the axes on which the earth spins, and are as distant as any two regions of the world can be. These two polar realms are, in fact, mirror images of one another. Antarctica, to the south, consists of a continental landmass surrounded by oceans, whilst the northern Arctic consists essentially of sea, ringed by the North American and Eurasian continents and their off-lying islands. The physical similarity of the poles lies in the fact that both are capped with ice, obliterating the features of land and sea, and both experience severe climatic conditions.

Antarctica, which accounts for almost ten per cent of the earth's land surface, is in fact the highest of the world's seven continents, rising an average of 1,830 metres above sea level. Its centre is a great plain, ringed by lofty mountains, some of which, such as Mount Erebus, are active volcanoes. The Antarctic Peninsula reaches out towards South America, and is thought to be a geological continuation of the Andean cordillera that became separated from the rest of the range many millions of years ago when the world's continents were assuming their present formation. The presence of vast coal reserves, which can only form in warm, humid conditions, suggests that the remainder of Antarctica was an equatorial region some 100-200 million years ago, having drifted slowly south since that time. The Peninsula is separated from this main Antarctic shield by the Transantarctic Mountains, many of the peaks of which exceed 4,000 metres.

Permanently swathed in a sheet of ice that is, on average, 2,450 metres thick, little of this topography has been exposed to the elements for over fifteen million years. Of an estimated fifteen million square kilometres of land, less than one per cent is free of this mantle – this percentage covers only a few coastal

1

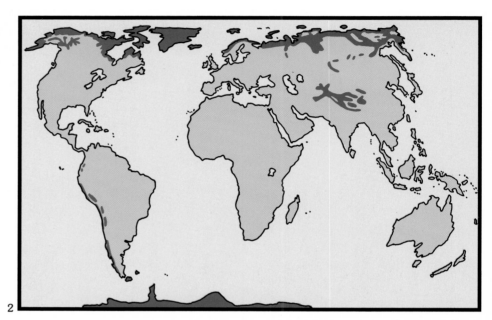

2

3

4

The world's polar habitats (**2**) are, naturally, largely confined to the northern and southern axes of the planet, with the limited addition of high montane tundra. Icebergs (**4**), arising from continental glaciers or oceanic pack ice, are a predominant feature of polar seascapes.

areas and mountain peaks. Some parts of the continent are crushed under the weight of up to five vertical kilometres of ice, and the total volume of the icecap is estimated at thirty million cubic kilometres.

This Antarctic icecap, which contains about three-quarters of the world's fresh water, spreads outwards from the centre. Where it reaches the open ocean, the force of the waves underneath causes huge chunks to break off, forming icebergs. During the summer the ice retreats towards the coasts, warmed and melted by the sun's rays but, during the winter, it creeps outwards again to form an irregular circle north of latitude 60° south in places. Some scientists use this maximum extension of the icecap to define the boundary of Antarctica, but others suggest that the limit is better marked by the Antarctic Convergence, the point where the warmer seas of the Indian, Atlantic and Pacific oceans mix with the cold Antarctic waters.

On the opposite side of the globe lies an ocean of ice from whose shores radiate the continents of the northern hemisphere. As with the Antarctic, there is some disagreement about the actual extent of the Arctic. For many years it was considered to comprise all the land lying within the Arctic Circle, north of which there is at least one twenty-four-hour day and one twenty-four-hour night per year, but a more meaningful definition relies on the position of the 10°C summer isotherm. To the north of this imaginary line the average July air temperature does not exceed 10°C, which has a marked impact on the vegetation of this realm. As a consequence, the isotherm limit coincides almost exactly with the northern limits of the boreal forest, or taiga – the Arctic tree line.

The Arctic Ocean, some five times the area of the Mediterranean Sea, is smothered with solid ice that may be up to 3,000 metres thick in the central region,

5

while drifting sea ice covers much of the remainder. The surrounding landmasses are also icebound in their northernmost reaches – the Greenland icecap covers almost the entire subcontinent. As in the Antarctic, volcanic activity is not uncommon. Iceland is the youngest country in Europe, having risen from the sea just sixty million years ago; this land of ice and fire still trembles from the aftershock.

The Arctic can be divided into two main realms: icecap and tundra. The icecap is a barren landscape occupying the highest latitudes over the ocean. The word tundra is derived from the Finnish for 'open plain' and it occurs in a broad swathe surrounding the icecap on the peripheral land masses. The tundra, sandwiched

when the sun remains below the horizon, the North Pole becomes a nightmare region of continuous darkness, while throughout the remaining six months of the year, the summer, the sun never sets. A similar pattern of light and dark, but at alternate times of the year, occurs in Antarctica.

Six months without the sun is the main reason for the numbingly low winter temperatures suffered by the polar regions of the earth, and without its warming rays, life is unable to exist. Even during the summer the sun is always low in the sky, its rays striking the curved polar extremes only tangentially. This means that light from the sun has further to travel through the atmosphere so that, on reaching the ground, it has less

Flocks of pintado petrels (6), also known as Cape pigeons, wing across the polar seas in search of plankton, settling on the water surface to feed. Immature golden eagles (3) are frequently seen over the Alaskan glaciers.

6

between the permanent ice fields and the northern coniferous forest, the tundra is where the majority of Arctic wildlife is to be found thanks to the more favourable summer climate of this region. Uncovered by the retreating ice only 8-15,000 years ago, the tundra is the earth's youngest biogeographical region. Similar zones are also found in other parts of the world: some ten million square kilometres of alpine tundra occurs above the tree line of high mountains, and small areas of tundra exist on the subantarctic islands.

The tundra comprises a monotonous plateau, levelled by the weight of the ice sheets that have rolled back and forth over it during successive Ice Ages. Much of the surface is littered with debris left by the retreating glaciers, and it is snow-covered for most of the year. During the summer much of the snow melts, although a few centimetres below the surface of the ground the soils remain frozen. This frozen layer is appropriately known as permafrost and may be over one-and-a-half kilometres thick. The presence of this impermeable subsurface stratum means that the summer meltwaters cannot drain into the soil, creating large marshlands and swamps in low-lying areas.

The climatic similarity of the Arctic and Antarctic is a result of the earth's movement around the sun and the slight tilt of its axis. During the winter months,

7

The humpback whale (1), seen here in its summer feeding waters in Glacier Bay, Alaska, makes spectacular leaps from the ocean surface despite weighing over forty tonnes and measuring some eighteen metres when adult. The polar bear (7), one of the world's heaviest terrestrial carnivores, is undisputed lord of the Arctic coastal fringes, equally at home on land or in the water. Chinstrap penguins (5) breed on the Antarctic landmass itself as well as on the surrounding islands, often in colonies numbering millions.

warming potential. In addition, the rays are spread out over a greater surface area, and thus cannot warm the earth as efficiently as at the equator, where they enter the atmosphere almost perpendicularly. The icecaps themselves also act as mirrors, reflecting much of this solar energy back into the atmosphere.

The Arctic region has a less severe climate than its southern counterpart, largely due to the warmer sea and air currents emanating from Eurasia and North America during the summer, which cause the peripheral ice to melt, exposing land and sea to the heat of the sun. Antarctica, on the other hand, is entirely surrounded by vast expanses of sea, and the continents of the southern hemisphere are too far away to have any significant warming effect. This, combined with the substantial height of the Antarctic continent, means that summer air temperatures rarely rise above freezing point even on the outskirts of the icecap, while in midwinter minus sixty degrees celcius is about average over the South Pole. As a result, the southern icecap is several times as thick as its northern equivalent.

Although ground ice covers both poles, very little precipitation occurs. In Antarctica it never rains, whatever atmospheric moisture there is falling as snow, due to the year-round freezing temperatures.

Even so, many of the frequent blizzards encountered in Antartica are due to ground snow being swept up and driven before the howling winds, rather than to true precipitation. It has often been referred to as a 'white desert'; a very apt description considering that the highest level of Antarctic precipitation is only 500 millimetres annually. Any water resulting from melting snow or ice evaporates almost immediately in the violent winds, which can reach speeds of up to 160 kilometres per hour.

Despite being less severe than the Antarctic, the Arctic climate is by no means benevolent. Strong winds rake its plateaux at all times of the year except in high summer, and snowstorms are almost continuous during the winter. Much of the northern tundra region suffers winter temperatures of -50°C or less, although along the coasts of southern Greenland and Iceland the westerly ocean currents maintain the temperature at around -7°C. Annual precipitation may be as low as 250 millimetres, but the low rates of evaporation here, even in summer, mean that much of this water is available to plants and animals – unlike in hot desert regions with similar levels of rainfall. Even in the height of the Arctic summer, when the hottest regions may experience air temperatures of up to 27°C, soil

The lesser golden plover (**1**) of North America and Asia, slightly smaller than its European counterpart, breeds on the Alaskan tundra, scraping a shallow nesting depression in the ground. The yellow-brown mottled chicks (**2**) are well-camouflaged from predators amongst the lichens and dwarf shrubs of the Arctic heath. The southern giant petrel (**5**), which breeds on oceanic islands in Antarctica, supplements its fish diet with eggs and chicks from the nearby penguin colonies.

1

2

3

Southern elephant seals (**3**) breed on islands in the South Atlantic and southern Indian oceans as well as on the Valdés Peninsula in Argentina. The males have smaller noses than their northern counterparts and their bodies are more flexible, despite the fact that they are the largest of all pinnipeds, weighing up to four tonnes and measuring some four-and-a-half metres in length. Weddell's seals (**4**) live further south than any other mammal in the world. The adults spend long periods of time beneath the ice, gnawing special breathing holes with their strong, protruding canines.

The musk-ox (**6**) is so called because of the pungent scent emitted by the rutting bulls. Both sexes have flattened horns, used in the male both in defence of the herd and also in the violent fights which establish the order of dominance between rivals. Musk-oxen are today confined to the North American tundra, though until the hunting techniques of early man improved they were also widespread across northern Europe.

temperature seldom exceeds freezing point since most of the sun's energy is expended melting the snow and upper permafrost.

In summer, the seas surrounding the icecaps are much less saline than other oceans since continual streams of meltwater dilute the surface layers. However, the salinity of the water increases with depth. These cold, dilute seas are nonetheless highly productive, since constant ocean currents bring nutrients to the surface as well as exchanging saline for dilute waters. Moreover, cold waters hold more oxygen than warm ones. Light, an essential factor in any food chain, is abundantly available for some six months of the year, especially in subarctic and subantarctic waters which are largely free of ice in summer. Microscopic phytoplankton such as diatoms make full use of these light, oxygen-rich seas, and are subsequently consumed by zooplankton, which are unable to synthesise their own energy from the sun.

Zooplankton in turn provide food for a number of predators, ranging from larger zooplankton and fish to the great baleen whales. Perhaps the best-known marine microorganism of Antarctic waters is krill. These small, shrimp-like crustacea form the basis of several food chains in the southern oceans. They are consumed in large quantities by many creatures, especially fish, which are in turn eaten by marine birds and seals, including sea elephants. These vertebrates are then eaten by the advanced predators of the Antarctic, such as killer whales and leopard seals. It is undoubtedly the peripheral areas of the southern icecaps that hold the most animal life, all of which is ultimately dependent on the riches of the sea for its existence, rather than on the barren, icy land of Antarctica.

The situation is not quite as critical in the Arctic, since much solid land is revealed by the melting ice every spring. However, vast numbers of marine birds and seals are similarly reliant on the more prolific harvest of the seas. In the tundra, where true terrestrial ecosystems exist, the food chains are nevertheless very short, and all of them are dependent upon lichens at the lowest levels. The fact that these relationships are so uncomplicated makes them very sensitive and easily destroyed, since the species are highly interdependent.

The polar icecaps themselves are incapable of supporting higher plant life because water is not available to the roots of plants, soils are almost nonexistent and, above all, because temperatures are below freezing point for much of the year. It is not surprising, therefore, to find that the whole continental landmass of Antarctica supports just two species of flowering plant – a tussock grass and a cushion-forming member of the pink family – which grow only where the ice retreats sufficiently along the Antarctic Peninsula to reveal land. No flowering plants have ever been found south of the 70° line of latitude, but some snow-melt areas are colonised by brightly-coloured mosses and grey-green lichens during the summer.

1

The grass *Deschampsia antarctica* (**2**) is one of the few vascular plants able to grow on the Antarctic landmass, the vegetation being dominated by Usnea lichens (**4**) and mosses (**5**).

2

3

4

5

Over 400 species of lichens, as well as about seventy different types of moss, have been recorded on Antarctica. These are confined largely to the coastal regions where they may form carpets up to one metre thick.

Some of the subantarctic islands (such as Kerguelen and Macquarie) however, lying beyond the reach of the winter ice sheets, have a more benevolent climate which, in terms of the length of the growing season, availability of water and absence of permafrost, should permit trees to grow. But the only vegetation present on these islands grows in wet tussock grasslands, heaths and bogs; a landscape which resembles the Arctic tundra. Some researchers have suggested that constant winds are responsible for the lack of tree

6

Typical elements of the Arctic landscape include the arctic-alpine mountain avens (**1**), reindeer (**6**), known as caribou in North America, and Arctic hares (**3**), which, though solitary for the rest of the year, form large groups during the autumn and winter, possibly for greater protection against predators.

growth here. Many of these islands are volcanic in origin, and all are remote, inhospitable places. As a result of their isolation, such flora as exists is composed predominantly of endemic species – the Kerguelen cabbage and related species are, in fact, the sole representatives of an endemic genus.

The central Arctic region is climatically similar to Antarctica, but around the margins of the icecap, warmed by oceanic and atmospheric currents originating from Eurasia and North America, plant life can exist. Much in the same way as the driest desert will bloom at the onset of rains, so the tundra comes to life with the return of the sun. As the flowering season seldom lasts longer than six weeks, annual plants, which overwinter in seed form, are rarely encountered

the delicate buds at the centre of the plant, but also trap air to increase insulation and allow the maximum possible light to fall on the leaves in order to increase the rate of photosynthesis during the short growing season. These low-growing plant forms, along with such prostrate species as the Arctic willow, which may raise itself only centimetres from the ground but can be as long as five metres, are designed to reduce wind desiccation and snow blasting. Their proximity to the ground ensures that even low snowfall covers the plants, providing winter insulation and speeding up flowering, since temperatures are normally higher at ground level.

With moisture in the Arctic locked up in the form of ice, water is at a premium, and not readily available

Upland geese (7) breed in southern South America and on nearby subantarctic islands, whereas snow geese breed in the Arctic tundra, the goslings (10) leaving the nest a few hours after hatching.

7

8

9

10

since there is seldom sufficient time for them to develop, flower and set seed. Instead, as in high mountains, most of the plants that thrive in the Arctic tundra are perennial, reproducing vegetatively by means of runners and bulbils.

Many of the climatic adaptations displayed by tundra plants in their efforts to survive under such restrictive conditions are the same as those seen in certain mountain species such as cushion-forming moss campion, purple saxifrage and Arctic forget-me-not, plants such as spotted and prickly saxifrages, and many grass and sedge species which do not shed their dead leaves, but instead accumulate them around the living tissues, and rosette plants: thrift, poppies, and snow saxifrage. These arrangements not only protect

Whooper swans (8) breed right across northern Eurasia, although they favour freshwater lakes in the boreal forest more than the open tundra. Penguins are strictly southern-hemisphere birds, some species, such as the Adélie penguin (9), even nesting on Antarctica itself.

to plants. Hence many species are very hairy in order to trap a layer of still air around them and so reduce water loss by evaporation. These hairs also serve to insulate delicate flower buds and new leaves from the intense cold; woolly lousewort and boreal Jacob's ladder are good examples of this phenomenon. When winter returns and all water supplies are frozen, the staghorn lichen responds in much the same way as the desert resurrection plants: the whole plant curls into a ball to protect the fruiting bodies and rolls before the wind until the following spring, when it comes to rest and spreads out again.

Many Arctic species have evergreen leaves, enabling them to photosynthesise whenever sufficient light is available; the carbohydrates thus produced are often stored in underground rhizomes or fleshy roots, so that the plants can flower as soon as conditions are favourable. For the same reason, many plants produce their flowering buds at the end of the previous summer. As a response to the low temperatures, the leaves and flowers of many Arctic plants are darkly coloured in order to absorb the sun's heat more efficiently. Some have even evolved a mechanism whereby the flowers can turn to track the sun, thus ensuring that they always receive the maximum amount of light possible. Species such as the Arctic poppy and glacier avens have cup-shaped flowers so that sunlight is reflected from the petals down onto the reproductive parts of

2

the flower, thus accelerating seed development. Another function of this adaptation is that the warm interior of the flower attracts insects, ensuring that pollination and fertilisation take place.

The paucity of Antarctic plant life is reflected in the higher trophic levels of the terrestrial ecosystem. Although abundant, the birds and seals that line the shores during the summer are wholly dependent upon the sea for their food, venturing onto the coastal ice to breed; in fact, the only vertebrates that live permanently on the continental landmass are the people who man the Antarctic survey bases.

Thus the land-based food chain consists only of a few species of invertebrates, mostly wingless: springtails which feed on the lichens and mosses, and mites which prey on the springtails. Many of these species contain a chemical which acts as an antifreeze in their blood, or are able to dehydrate and shrivel up in severe weather so that ice crystals do not form in their tissues and rupture the cells. Some do not even try to survive during the long, cold winter, relying instead on resistant eggs, which lie dormant until favourable conditions return. Of fifty species of insects and mites which have been recorded in Antarctica, about half are parasitic, living on the birds and mammals of the coast. Most of the remainder live amid the lichens that thrive on the guano produced by seabird colonies.

These seabird colonies are probably the largest in the world. Every year some 100 million birds breed on the shores and islands of Antarctica, gorging themselves and their young on the abundance of marine life during

1

3

4

the continual daylight of summer, and then migrating when the nights start to lengthen into the days, so avoiding the harsh winter.

The penguin is considered to be the most typical bird of the Antarctic, but the existence of equatorial species, such as the Galápagos penguin, prompted researchers to look more closely at the group. In fact their bodies are designed specifically for an aquatic way of life, rather than to cope with the severe cold of the Antarctic: their wings have become flippers and their bodies are streamlined, their tails act as rudders, their feet are webbed and their feathers form a dense, waterproof coat. Only the thick layer of subcutaneous fat may be of assistance to the penguin in the freezing

The best-known inhabitants of Antarctica are undoubtedly the penguins, flightless birds in formal black and white plumage. Among the few species that can weather the severe climate of the Antarctic landmass are king (**3,4,9**) and chinstrap (**2**) penguins.

5

polar climate. It is now thought that most penguins are able to tolerate a wide range of climatic conditions, but the majority have favoured Antarctica for breeding and moulting – the only two activities for which they need to be on land – because of the relative absence of terrestrial predators there.

Only emperor and Adélie penguins are exclusive to Antarctica, but five other species regularly breed within the maximum limit of the winter ice: gentoo, chinstrap, macaroni, king and rockhopper. Many of the largest colonies are on the surrounding islands: fourteen million pairs of chinstrap penguins breed on Zavodovski Island (in the South Sandwich group), which is only six kilometres across; half a million

6

The Arctic tern (**6**), apart from nesting as far north as any other bird, also has one of the longest annual migrations. The breeding distribution of the lesser snow goose extends from Arctic North America across the Bering Straits into Siberia, although the distinct blue phase (**1**) is found only in the eastern part of this range. Short-eared owls (**8**) are ideally suited to life in the tundra since they are ground-nesters and take advantage of the abundance of voles, their preferred quarry.

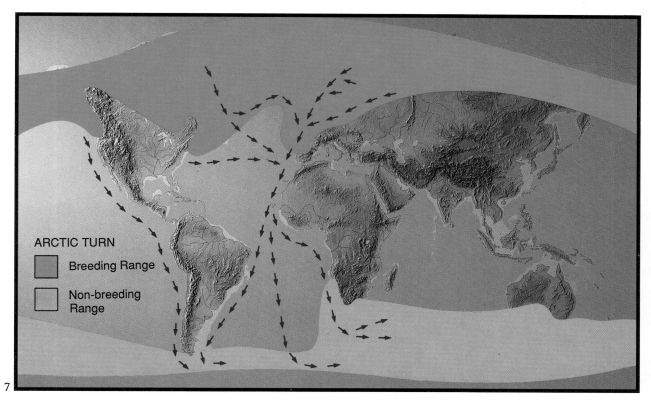

ARCTIC TURN

■ Breeding Range

□ Non-breeding Range

7

8

9

10

rockhopper penguins nest on Nightingale Island, and colonies of up to 100,000 emperor penguins are found on Coulman Island, close to the coast of Victoria Land.

At the time when most birds are migrating north to avoid the winter, the emperor penguins are heading south into the gathering darkness. Because this is the largest of all penguins, the young take longer to develop; the process of egg incubation must start earlier if the chicks are to be full grown and able to fend for themselves by the following winter. Once the eggs have been laid, the females embark on the long trek back to the sea, whilst the males huddle together in groups of several thousand, a single egg balanced on

their feet and tucked under a loose flap of abdominal skin to keep it warm.

The temperature at the icy nesting site may drop to -40°C and winds may exceed 140 kilometres per hour, but the emperor penguin, of all the world's birds, is able to endure the coldest conditions. The females spend the winter fattening themselves at sea, timing their return for the hatching of the chick. During this time the males eat nothing, losing up to half their body weight, but when the chick hatches the female takes over the care of the young and the male heads out to sea to replenish his depleted body reserves. When the chicks are large enough, the long march to the coast commences. There they learn to catch food, mainly

Penguins breed in flat coastal regions in enormous colonies where the general confusion is such that a passing sheathbill is able to take advantage of a stray egg (**10**). Blue-eyed shags (**5**) prefer to nest on narrow ledges in the cliffs.

fish and large squid, and store up sufficient food reserves for the coming winter. Nevertheless, even in a good year, only about forty per cent of emperor penguin chicks reach maturity.

Generally speaking, all the birds that breed in the Antarctic are seabirds – although South Georgia boasts a species of pipit and an endemic subspecies of the Chilean pintail duck. Despite being seabirds, however, not all of them obtain their food directly from the sea. The great skua, occupying the niche of top predator in the terrestrial food chain, feeds largely on penguin eggs and chicks, or robs other birds of their catch. It also eats fish if other food is scarce. The sheathbills, of which there are two species – the wattled and lesser – are the only birds of the Antarctic that do not possess webbed feet; they are thought to be the modern descendants of the common ancestor of all seabirds. These birds are essentially scavengers and carrion eaters, although they will eat the eggs and young of seabirds, including those of other sheathbills, and may even finish off an injured adult penguin. Both skuas and sheathbills are opportunistic feeders and have learnt to take advantage of the rich pickings to be had around the Antarctic research stations.

Antarctic breeding birds generally consist of a few species represented in vast numbers. Some of those that are directly dependent on the riches of the sea include blue-eyed shags, which make use of sheer cliffs as nesting grounds, giant petrels, which scrape shallow nests on the plateau areas, and several species of albatross, of which the wanderer has a three-and-a-

1

The rock ptarmigan (**3**) has a truly circumpolar distribution in the Arctic and is one of the few birds to remain in this harsh environment throughout the winter. Like many members of the grouse family, ptarmigan undergo a series of moults so that their plumage matches their environment, ranging from pure white in the winter to brown or grey in the summer. Blue-eyed shags (**2**) show several physical characteristics typical of birds that spend most of their time in the water – legs positioned towards the rear of the body, powerful webbed feet and small wings.

2

3

half-metre wingspan – the largest of all seabirds – and glides effortlessly over the oceans in search of food. It expends less energy in the process than any other flying creature. But it is the Arctic tern that travels further than any other living creature in the course of the year. This slender, red-billed tern breeds on the other side of the world; on the shores of the Arctic Ocean – within a few hundred kilometres of the North ole – but spends the other half of the year in Antarctica. In order to do so it migrates over 40,000 kilometres per year, flying for about eight months of the year, and is the only wild creature in the world to experience two summers per year.

The shores of the continents that fringe the Arctic Ocean and the numerous islands of this region are populated by seabirds that are not unlike the southern penguins. Indeed, the great auk, a large, flightless bird, hunted to extinction by the middle of the nineteenth century, was originally called 'penguin'; this name was later inherited by the similar-looking birds in the southern hemisphere. Living members of the auk family include razorbills, guillemots, puffins and the diminutive little auks themselves, and these birds resemble present-day penguins in many ways. They are mostly black and white and have an upright stance

4

on land, but ride the waves in a horizontal position (facilitated by feet placed well to the rear of the body) and are excellent underwater swimmers – by which means they obtain most of their food. Like penguins, they breed in huge, noisy colonies, favouring precipitous rock faces and inaccessible rock stacks since the Arctic is home to several voracious terrestrial and shoreline predators. Unlike penguins, however, they have retained their power of flight, although their short, broad wings – so useful beneath the water – are rather inefficient in the air, producing the 'whirring' sound in flight that is so characteristic of this family. Like the summering wildfowl and passerines, auks also tend to migrate southwards with the onset of winter.

Emperor penguins (**1**), the largest of all species, and their close relatives the king penguins (**6**), both have distinctive yellow-orange ear markings. Adélie penguins (**7**) find traversing land on foot an arduous process and instead may opt to 'toboggan'.

6

Gulls also take advantage of the richness of Arctic seas during the summer. Ross's and ivory gulls are true polar species, wintering on the Arctic shores or around the edges of the pack ice, and some glaucous gulls spend the winter on the coasts of Greenland. However, most of the gulls that breed in the Arctic, including Sabine's and Iceland gulls, and kittiwakes, spend the winter in warmer climes. Some gulls are able to roost on the ice for long periods despite the fact that their feet have no apparent insulation. In fact, the internal foot temperature may be as low as 0°C, although the rest of the body is at a comfortable 38°C. The network of blood vessels in their feet and legs operates as a heat-exchange mechanism, so that only cool blood circulates through the feet, thus avoiding unnecessary heat loss.

The Arctic tundra possesses a much more diverse summer avifauna than its southern counterpart, providing a summer home for more than a hundred breeding bird species, including swans, geese, ducks, various wading birds and passerines. All of these flock in their thousands to the tundra lakes and marshes that form over the permafrost during summer. Here they nest and rear their young, feeding on the swarms of summer insects and the lush vegetation of the wetlands,

7

which are bathed in almost perpetual sunlight. These Arctic insects are similar to high alpine forms in their adaptations to climate. Many are melanistic, their dark colour increasing the absorption of the sun's warming rays, and many have dense hairs to retain this heat within the body. Most are wingless or have reduced wings in order to cut down on heat loss over such large surface areas, and some synthesise glycerol in their body fluids to prevent the formation of damaging ice crystals at low temperatures. Others are able to

Like many tube-noses, both black-browed (**4**) and sooty (**5**) albatrosses perform an elaborate courtship display. Each pair will rear a single voracious chick.

reproduce parthenogenetically – without the eggs being fertilized – the males of these species are thus rare or nonexistent, and most of the insects are only active in good weather.

Birds of prey also arrive in the summer to take advantage of the large numbers of young birds and rodents that the tundra yields at this time of year. The merlin migrates south to the Mediterranean area in the winter, but returns in March to breed in the transitional tundra/taiga zone, feeding mostly on winged prey – large insects and small birds – but sometimes supplementing its diet with a few lemmings. Rough-legged buzzards and short-eared owls are also found in the tundra during the summer, although these are more wide-ranging species, being found across the whole of the northern hemisphere. As in the Antarctic, skuas also fill a predatory niche, although in the Arctic the great skua is less common than the related Arctic, pomarine and long-tailed skuas.

The shortening of the days prompts these Arctic breeding birds to undertake the long migration to their wintering grounds in the south, since the absence of insects and the disappearance of the vegetation beneath a thick blanket of snow and ice make it impossible for such species to survive the Arctic winter. But others remain in the tundra all year round. Some, such as the willow grouse and ptarmigan, are well-equipped for the winter cold, having thickly feathered legs and feet, both for insulation and to facilitate travel across the surface of the snow. Both change their plumage from speckled browns and greys to glossy white during the autumn, thus becoming almost invisible against the snow; the density of these winter feathers is also higher. Small birds, such as the Arctic redpoll, forage beneath the insulating snow, having a high metabolic capacity in order to maintain their body temperature. Predatory and scavenging birds – ravens, peregrine falcons, gyrfalcons and snowy owls – are also able to survive in winter as there is no shortage of their main food items: small rodents that do not hibernate, as well as hares and grouse.

Tengmalm's and pygmy owls, and mammals – lemmings, voles and hares – sometimes even ermine. The rich diet of both these birds of prey allows them to expend energy maintaining their body temperature, and also enables them to hunt more effectively.

The tundra of the northern polar region is also home to a variety of mammals – especially during the summer – ranging from tiny lemmings to huge musk ox. In fact, higher vertebrates are favoured in this realm by the purity of the air, the almost total absence of infectious diseases, and a low density of human population. The twelve species of lemming that live in the circumpolar region provide the basic food of many small predators. In cycles of three to four years (some researchers identify cycles up to three times as long) their populations explode in response to favourable conditions. At such times the breeding rate over the six-month summer period is some five litters per year, with between two and eleven young per litter. When

The national symbol of the United States of America, the bald eagle (**2**) frequents the northern wilderness areas, but cannot be regarded as a true Arctic predator to the same extent as the snowy owl (**10**). Trumpeter swans (**1,3**) once ranged over a large area, but are now confined to a few lakeland reserves in northwest North America, especially Alaska.

1

2

3

Like grouse, snowy owls have well-covered legs and feet, and also have copious and compact feathers all over their body. The male is completely white, but the female is barred with black or grey on the head, wings and breast to act as camouflage when she is incubating her eggs, usually in a hollow excavated amid lichens. Unlike most members of the family, the northernmost snowy owls are forced to hunt during daylight for the summer months of the year, when the sun does not drop below the horizon. Their preferred prey is lemmings and the various species of hare – snowshoe, varying and Arctic – which inhabit the tundra. Gyrfalcons are the largest of the world's falcons, varying in colour from grey to almost pure white. They feed both on birds, including other predators, such as

4

The black-throated diver (**8**) is a typical tundra breeder, but if spring comes late and the ice does not thaw it will not breed that year. Antarctic skuas (**4**), like albatrosses, have an elaborate and noisy courtship ritual.

5

6

7

8

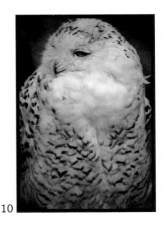

9

10

all available food supplies are exhausted, the apocryphal 'migrations' of these lemmings take place, although the legendary vast hoards are now thought to owe more to fiction than fact.

Lemmings, shrews and voles do not hibernate during the winter. Instead they rely on the insulating blanket of snow to protect them from the inclement weather. However, in isolated burrows, the Alaska marmot and Arctic ground squirrel enter a state of true hibernation, lowering both their body temperature and metabolic rate. The grizzly bears of the southern reaches of the tundra become dormant – a process in which the body temperature remains normal, distinguishing it from true hibernation. As in the high mountains of the world, those species that remain active above the snows during the winter – such as ermine, snowy owls, ptarmigan and varying, Arctic and snowshoe hares – usually change colour to blend in with their surroundings.

The two most conspicuous large herbivores of the tundra are the caribou and the musk ox. The North American caribou is in fact the same species as the Eurasian reindeer, but the latter now exist only in semi-domesticated herds. They are gregarious beasts that undertake long journeys from the shelter of the boreal forests during the winter to give birth on the open tundra during the summer. Both sexes have deciduous antlers, which appear to play a part in the social hierarchy of the herd. The males lose their antlers in November, after the rut, but the females keep theirs well into the winter, during which time they appear to take over the leadership of the herd.

Caribou are ideally suited to the tundra habitat. Their thick coats are composed of hollow hairs which trap air and have excellent heat-retaining properties. Their broad, splayed hooves have concave soles and sharp edges to provide good traction and prevent sinking on icy snow and marshlands alike, as well as being useful in clearing compacted snow to expose the layer of lichens - their main winter food. On their long seasonal journeys, caribou travel in closely-knit groups at speeds of up to thirty kilometres per hour to deter attacks by their major predator: the wolf. Although caribou herds declined dramatically as a result of overhunting around the middle of the century, more sensible culling and protective legislation in Canada have promoted a recovery in numbers in recent years. promoted a recovery in numbers in recent years.

The musk ox resembles the bison of the North American prairies in appearance, but its closest relative is, in fact, the takin of the Himalayas. Unlike the caribou they do not have fixed migrations, although they may seek shelter during the winter. Their pelt is thick and fleecy, each hair being up to thirty centimetres

1

Musk-oxen, Arctic foxes, polar bears and walruses are all virtually confined to the Arctic biome. The musk-ox (**1**) is the only ruminant to have successfully adapted to life in such harsh conditions, remaining even during the severe winters.

The Distribution Range of the Walrus and the Polar Bear.

■ **Polar Bear Range**

■ **Distribution of the Walrus**

3

4

2

long, and they store huge fat reserves under the skin in order to survive the winter, when most of the vegetation is buried under the snow. The males may weigh over 400 kilogrammes and are armed with a fearsome pair of horns, which almost meet across the brow. When threatened by danger, the small herds form a ring – females and young in the centre and males facing outwards: a defence which has proved to be worse than useless against the rifles of the tundra trappers.

The most spectacular terrestrial predator of the northern icecap is undoubtedly the polar bear. It is so well-adapted to this frozen land that it has been called the 'lord of the Arctic'; indeed, the word Arctic is derived from the Greek word *arctos* meaning bear. Researchers think that it is a relatively young species, having become separate from the brown bear only during the last Ice Age. The polar bear is equally at home on both land and sea, feeding largely on ringed and bearded seals and young walruses, however, it will also feed off carrion. Along with the Kodiak bears of Alaska, it is the world's largest predator, weighing up to 800 kilogrammes. This white bear's only natural enemy is the killer whale, but since the seventeenth century the polar bear has also suffered the onslaught of over-hunting for its superb pelt. Following a dramatic decline in its numbers, a circumpolar ban on hunting has resulted in the resurgence of this species; today

5

During the winter the Arctic fox (**9**) displays two distinct colour variations, blue and white, while its summer coat is brown or grey. Lemmings are its main prey, though individuals living close to the coast may take fish and carrion from polar bear kills.

there are some 25,000 polar bears in the U.S.S.R. alone.

A denizen of the Arctic coast and sea ice, the polar bear is an expert swimmer, capable of reaching speeds of eight kilometres per hour, and is even able to sleep in the water. On land its short, strong claws and furry-soled feet enable it to traverse the ice with ease, facilitated by the fact that, like man, it is a plantigrade animal. In the absence of caves, gravid females normally excavate dens in the snow, in which to give birth out of reach of the biting winds and winter blizzards.

A second predator that is more or less confined to the northern polar region is the Arctic fox, but this canine is almost completely terrestrial in its habits. Its diet is composed mainly of carrion, but it prefers live prey so that, in years when lemmings breed prolifically, it changes its feeding habits accordingly. The Arctic fox has a phenomenal breeding rate: the females give birth to as many as twenty-five cubs per year, although most of these do not survive more than six months. Even mature foxes rarely live for more than a few years in the harsh Arctic climate, despite being able to remain active at temperatures of around -50°C without a corresponding increase in their metabolic rate. There are two variants of the Arctic fox: a brown one which moults to become white in the winter, and a dark blue one which becomes paler at the onset of the snows. Like the polar bear, its number have declined

due to the huge demand for its soft, luxuriant coat, although the problem has been largely alleviated by the establishment of fur farms.

In contrast, the only mammals that can survive in Antarctica are dependent upon the sea for their survival. Perhaps the best-known and best-adapted to these cold waters are the seals. Of thirty-four species in the world, eight are found in these southerly waters: five true seals (Weddell, Ross, crabeater, leopard and the southern elephant), two fur seals (Antarctic and subantarctic), and Hooker's sealion. True seals, including elephant seals, are thought to have evolved

Polar bears (**3,5,7**) are widely distributed across the whole circumpolar region, but at very low densities. They prey mainly on bearded and ringed seals taken when they come up to breathe, though young walruses may also fall victim. There are two distinct subspecies of walrus, one living in the North Atlantic and the other (**4,6,8**) in the North Pacific. The adult males of the latter are larger, weighing up to 1,600 kilogrammes, and their ivory tusks may be up to a metre in length. Those of the Atlantic race rarely exceed thirty-five centimetres.

from an otter-like ancestor, whereas fur seals and sealions, known collectively as eared seals, are probably the descendants of a bear-like creature. The main visible difference between the two families is that true seals use their rear flippers for swimming, hunching like caterpillars to move about on land, whilst eared seals use all four flippers for 'walking' and the front pair for swimming.

The rarest of the Antarctic seals is Hooker's sealion, also known as the New Zealand sealion, which was nearly wiped out in the nineteenth century, but following protection measures, the group now comprises about 5,000 individuals. The largest, without a doubt, is the elephant seal, the bulls of which measure four and a half metres from end to end and weigh almost four tonnes. Not only are they the largest members of the seal family, they are the also the largest mammals to frequent terrestrial habitats in the world today.

Killer whales (**1**) have the longest dorsal fin of any whale, but the humpback whale (**3**) has the longest flippers, although it is better-known for its haunting voice. The isolated north Pacific and north Atlantic populations of humpbacks possess similar songs, but whales of one subspecies do not respond to recordings of the other.

through the ice that periodically freezes over their breathing holes, but they are able to remain submerged for more than an hour at a time. The crabeater seal is the most numerous member of this family in the Antarctic, feeding, despite its name, on krill which it filters from the water through its specially-adapted canines and incisors. It is especially active during the night, when these zooplankton rise to the surface waters of the ocean. There are some 150,000 voracious leopard seals in Antarctica and these feed on fish, seals and penguins, visiting the coastal fringes frequented by seals and chasing Adélie penguins under the ice floes.

Seals are also a characteristic element of the Arctic fauna. Northern elephant seals are a little less bulky

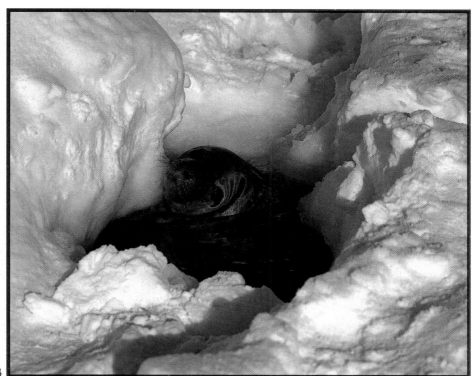

The Weddell seal is the most southerly of all living mammals, supremely designed for life in these freezing waters. In order to spend long periods underwater in search of food, its blood has five times the oxygen-carrying capacity of human blood, and it also possesses twice as much blood per unit of body weight as man. Only four centimetres below the skin, regardless of the external temperature, the body tissues are maintained at a constant 35°C, largely due to a thick, subcutaneous layer of insulating fat known as blubber, which is a common feature of all seals.

Weddell seals remain in the Antarctic all year round, spending most of their time in the water under the insulating ice sheets to avoid the extreme cold. They have larger than average eyes in order to see in the dim world beneath the ice, and, in very dark conditions, may navigate using echolocation. Their front teeth protrude markedly so that they can gnaw

than their southern counterpart, the males weighing a mere two and a half tonnes, although they may be up to seven times as large as the females. On the other hand, the inflatable noses of the males, from which these seals get their name, are better developed in the northern species. Research has shown that the northern, or Alaska, fur seal is able to dive to incredible depths of nearly 900 metres in search of food. The true seals of the Arctic are much smaller than the southern members of the family; they include the larga, bearded, harp, hooded or bladdernose, ringed and common or harbour seals.

Young northern elephant seals (**2**) show little sign of the pronounced 'nose' which characterises the adult bull. Weddell seal pups (**4**) take advantage of the breathing holes gnawed in the ice by their parents. Of five subspecies of harbour or common seal (**5,6**) only two frequent Arctic waters, where they give birth in July, about a month later than their temperate relatives.

But the most renowned seal denizen of the north is the walrus. Both sexes have long ivory tusks, essentially upper canines, which grow throughout the life of the animal; in the male they may be up to one metre long. They prove useful for hauling the walrus out of the water, much in the same way as a mountaineer uses an icepick, and may also be used for defence against their few enemies – polar bears, the occasional killer whale and, inevitably, man. Their main purpose, however, would seem to be for uprooting shellfish and sea urchins from the sea floor, located by means of sensitive vibrissae, or tough whiskers, which decorate the front of the walrus' face, between nose and mouth. Scientists are still not certain which seal family the

1

Northern elephant seals (**4**) are smaller than the southern species (**5**), although the males have longer proboscises. The northern elephant seal was hunted to the brink of extinction at the beginning of this century, but the last few hundred recovered dramatically and the population now numbers around 100,000. Weddell seal pups (**1**), born further south than any other mammal in the world, have a thick layer of blubber beneath the skin to protect them from the elements.

2

3

4

walrus is more closely allied to, since it possesses characteristics of both seal families: the eared (or *Otariidae*) and the earless (*Phociade*). Like most of the Arctic seals, it migrates south for the winter, often hitching a ride on a passing ice floe.

The abundance of krill in the Antarctic, as well as providing food for numerous seabirds and the crabeater seals, is also responsible for the highest concentration of whales in the world, including the largest mammal ever to have lived on this planet – the blue whale, which may reach a maximum length of thirty-three-and-a-half metres and weight up to 140 tonnes. Like the blue whale, most of the Antarctic species are baleen whales, possessing fringed plates suspended from the upper jaw with which to filter plankton, and two blow holes rather than the single orifice of the toothed whales. There are ten species of baleen whale, of which the sei, fin, minke, southern right and humpback are to be found in Antarctic waters. Like the majority of seals and seabirds, these whales are present in the Antarctic

5

only from February to May – although, over the course of the year, their gradual migrations span the globe, from the northern oceans to the southern seas. The world population of baleen whales today is estimated to be less than ten percent of what it was at the beginning of this century, having been devastated by whaling activities.

The only toothed whales that are regularly present in the Antarctic ocean are the killer whale, or orca, and the enormous sperm whale. These are highly predatory, their prey including fish, squid, penguins, seals and

Pilot whales (**2**), also known as blackfish, differ from their much larger relatives, killer whales (**7**), in having shorter dorsal fins, white patches confined to the belly and prominent, bulging foreheads.

even other whales. Although it has a worldwide distribution, the orca prefers cooler waters and so is most abundant at the poles. This handsome, piebald whale is easily identified at sea by its prominent dorsal fin, which may be six metres high. Although baleen whales are predominant in the Antarctic, the northern seas are undoubtedly the domain of the smaller toothed whales, including such bizarre species as the narwhal,

6

with its long, spiral tusk. The males of this species are the only naturally one-horned animals in the world. The beluga whale is another typical Arctic whale. It is almost white in colour and therefore well-camouflaged against the drifting pack ice. Hunting has taken its toll on the whale stocks of the north. The Bering Sea population of bowhead whales shrank by ninety per cent between 1848 and 1914 as the result of Yankee whalers operating in the area; their population is today estimated at less than 3,000 individuals, compared with some 75,000 two hundred years ago.

The polar regions are probably the least known regions of our earth; Antarctica was the last continent to be discovered and is still relatively unexplored. But as man increasingly expands into remote parts of the world, even these last wilderness areas are being threatened. The vast mineral resources of the Arctic are already being exploited, and the huge machines involved in the process of extraction leave vicious scars across the slow-growing tundra vegetation. The resources of the southern polar realm are currently protected by the 1961 Antarctic Treaty, but in a world desperately short of raw materials, how long this protection will last nobody knows.

Fur seals (3) belong to the group of pinnipeds known as eared seals on account of their external earflaps. They are basically divided into northern, or Alaskan, fur seals and seven species of southern fur seal, distinguished mainly on the basis of geographical distribution, such as those found on South Georgia in the Antarctic (6). All species have a thick, soft undercoat which is highly prized by the fur trade, a feature which distinguishes them from sea lions.

7

8

Atomic testing was a frequent occurrence in the Arctic, and the resultant radioactive particles were absorbed by lichens and thence into the higher levels of the food chain, with, as yet, unknown consequences.

Global warming and the recently discovered 'holes' in the ozone layer are likely to cause significant melting of the polar icecaps; it has been estimated that if Antarctica were to release all of its stored water reserves then the sea level would rise by some fifty-five metres, flooding many of the world's coastal cities.

However, it has been realised that man cannot continue to disregard the natural order of the world's ecosystems, and environmental groups are calling for the declaration of Antarctica as a world park – the only continent on earth still in a completely natural state. There may be some hope yet for the unsullied whiteness of Antarctica and its teeming wildlife.

In the Arctic, common seals (8) are not infrequently found in the larger freshwater lakes, having made their way up-river from the sea. These inland populations appear to be fairly settled, the seals feeding mainly on northern lake trout rather than the Arctic cod and whitefish diet of the marine animals.

· CHAPTER 4 ·

BOREAL FORESTS

The boreal forest, one of the most extensive vegetation types in the world today, is dominated by conifers, some of the most ancient and primitive tree species in existence. Encircling the earth like a verdant diadem, these coniferous forests separate the harsh plains of the tundra from the rich deciduous forests of the northern hemisphere. Although at first glance they appear to be dark, forbidding places, snowbound for much of the year, these forests harbour many birds and animals that welcome the isolation of a remote wilderness into which man seldom ventures.

Right: Tengmalm's owl, (top) lichen and (above) a muskrat.

The boreal, or northern coniferous, forests encircle the globe in a more or less continuous belt, interrupted only by the Atlantic Ocean and the Bering Straits. This 'green crown' is some 800 kilometres wide, on average, and measures no less than 10,000 kilometres from one end to the other, traversing Canada, the United States, Europe, Siberia and China; it is one of the most extensive vegetation types in the world. In Asia, these boreal forests are also known as *taiga*, from the Russian for 'dark and mysterious woodland'.

The boreal forests are sandwiched between the Arctic tundra to the north and the temperate deciduous forests to the south, a region lying roughly between the latitudes of 55° and 65°. The terrain has been shaped by long periods of glaciation during the Ice Ages; it is a low, flat landscape, scoured by ice, with scattered glacial lakes, streams and extensive sphagnum bogs, known as muskegs. Although the northern demarcation between tundra and coniferous forest is often sharp, the cutoff point to the south is less abrupt, grading into a broad band of mixed coniferous and deciduous woods, the mix governed for the most part by local soil conditions.

Conifers are members of the *Gymnospermae*, an ancient group of plants which first appeared some 300 million years ago in the Carboniferous period. These primitive gymnosperms increased their range and diversity of life forms during the Permian period and the early Mesozoic era, but by the Cretaceous period some species, such as the ginkgos, or maidenhair

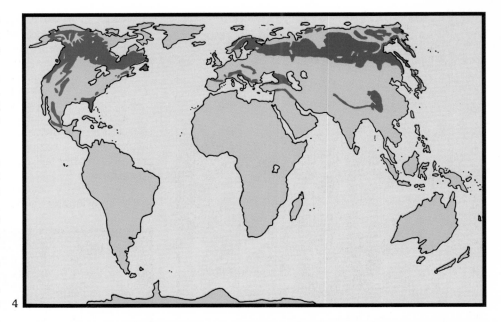

4

1

2

5

3

6

trees, and the cycads, were already starting to decline and they have few modern representatives. Nevertheless, conifers as a group persisted; they differ from most flowering plants in having naked seeds, that is, these are not enclosed in a protective case, but borne on scales which are then clumped together to form cones.

Coniferous forests are dark, monotonous landscapes. The close-growing trees allow little sunlight to penetrate as far as the forest floor, which as a result may be bare of all plant life, carpeted only in a thick, springy layer of fallen needles. However, more often than not, mosses and lichens flourish in even the darkest boreal forests, whilst the penetration of a little

The distribution of boreal forests (**4**) is clearly a northern one, although many montane regions further south support similar stands of coniferous woodland. The Scots pine (**3**) is one of the commonest forest trees of the European boreal forests, at its most spectacular in the ancient Caledonian pinewoods of Scotland, where trees up to fifty metres high are not unusual. In North America, however, the coast redwoods (**6**) which grow in the montane mist belt along the Pacific coast are probably the tallest trees in the world, reaching a maximum height of 120 metres in their natural habitat.

normal plant growth is impossible. Where this growing season constitutes less than six months of the year and, more particularly, where the frost-free period is less than four months, deciduous trees are usually replaced by coniferous species. This is because the average deciduous tree requires temperatures of more than 10°C for about 120 days of the year, whilst conifers can survive with less than thirty such days. Under such conditions tree growth is extremely slow, with the result that a hundred-year-old conifer may be only one-and-a-half metres tall and seven centimetres in diameter at the base.

Precipitation is low, usually between 375 millimetres and 750 millimetres annually, and much of this occurs

Predators of these dark forests on both sides of the Atlantic range from the northern goshawk (**7**) to the grey wolf (**1**), which has the greatest range of any living land mammal in the world, excluding man.

The North American black bear (**9**) is partially carnivorous, although vegetable foodstuffs – nuts, tubers and berries – make up a large part of its diet. Wild boar (**8**) are found in a wide range of habitats, but their compact bodies and short, strong legs are particularly well adapted to the harsh winters of the northern forests. Their young and those of the mule deer (**2**) of western North America frequently fall victim to wolves. The beaver (**5**) takes steps to protect its offspring when constructing its cone-shaped lodge surrounded by water. Even if a predator should enter the lodge, the beaver family is still able to escape via one of several underwater entrances, which also provide access when the lake is frozen.

more light will permit the growth of such attractive herbs as large-flowered wintergreen and of shade-tolerant woody shrubs, such as ground dogwood, baneberry and blueberry.

Weather conditions are far from benevolent in this northern land, with winter temperatures sometimes dropping as low as -70°C. In such northerly latitudes the summers are very short, but the days lengthen dramatically towards midsummer, in particular close to the Arctic circle. The higher temperatures at this time of year may result in an annual temperature range of over 100°C.

The growing season may be defined as that period when temperatures exceed 5°C, since below this

as snow. However, as evaporation is also low, even in summer, the climate is moist. Moreover, the trees themselves modify the climate, creating sheltered conditions where the air is still and humidity may reach 100 per cent. During the winter, owing to the fact that the soil may be frozen to a depth of two metres, most water is 'locked up' as ice and is thus not available to the trees for photosynthesis and other biochemical processes. In the summer, the abundance of meltwater creates a downward soil water movement, causing leaching of the free-draining substratum and the formation of extremely poor, acid soils, known as podsols.

Moving northward from the equator, climatic

conditions eventually become so severe that even conifers are unable to survive; this point is known as the Arctic tree line. It is not, however, a static boundary. In some places the forest is in retreat, due to fire, logging and damage by large herbivores, giving way to the even less hospitable tundra habitat. Elsewhere the conifers are still advancing to the north, following the retreat of the ice sheets from the last Ice Age, and have not yet reached their environmental limit. Along much of this boundary between forest and tundra there is a transitional zone, with scattered, dwarf trees and a ground layer composed of more light-demanding species such as tough grasses, cowberry and bilberry, all growing in a springy carpet of the lichen known as reindeer moss.

In Europe, during the ice ages, glaciers moving southward from the polar icecap trapped the temperate forests against the east-west mountain ranges of the Alps, Pyrenees and Carpathians, preventing their migrating southward away from the ice sheets. Many tree species could not survive the freezing climate which accompanied these ice sheets and thus became extinct. Although these forests later expanded

1

2

3

4

northward again at the end of the last Ice Age, the diversity of tree species was much reduced; Europe contains only about 1,000 tree species today. In America, however, the north-south alignment of the major mountain ranges permitted the forests to move southward with the shifting climatic belts and, as a result, present-day North American forests boast nearly three times as many tree species as Europe. Almost all of the temperate tree species present in Eurasia before the ice ages can still be found in China, which did not suffer extensive glaciation at this time.

In the North American boreal forests, where average January temperatures are about -20°C and only two-and-a-half months of the year are frost-free, the dominant trees are the white spruce and balsam fir on free-draining soils, and the lodgepole pine and sitka spruce elsewhere. The jack pine becomes dominant where fires are frequent, whilst more waterlogged soils support the tamarack and black spruce. Several deciduous trees are also associated with these conifers, in particular the paper birch, quaking aspen and balsam poplar. The humid coniferous woods of Alaska's Panhandle region have an especially rich ground flora, containing calypso and red coralroot orchids, single-

5

Although little sunlight reaches the floor of the northern boreal forest, irises (**4**) and the superb calypso or fairy slipper orchid (**3**) are able to thrive. The edible velvet shank fungus (**1**) and the succulent fruits of the salmonberry (**7**) are among the autumnal sources of food for the animals of these woodlands. Juniper (**5**), which grows from the equator to the Arctic, is commonly found in the drier coniferous woodlands of the Mediterranean region, while Eastern skunk cabbage (**2**) grows where the boreal forest meets areas of swampy grassland.

Two migratory species of the boreal forests are the hobby (**6**) in Eurasia and the monarch butterfly (**8,9**) in North America. To avoid the harsh winters, the adult monarchs travel over 3,000 kilometres south to coastal pine and cypress groves, often roosting in groups of more than 100,000.

6

7

8

flowered wintergreen, twinflower, yellow skunk cabbage and Indian pipe, also known as yellow bird's-nest.

The European coniferous forests, by contrast, are dominated by fewer species, notably the Scots pine on the dry, sandy podsols and the Norway spruce on the moister, richer soils. The pine woods of the Scottish Highlands are the sole remnants of the native coniferous forests that once covered the whole of the British Isles. The only conifer present is the Scots pine, often found in association with birch, rowan and juniper, with a ground layer of ling, bilberry and cowberry and such typical boreal herbs as chickweed wintergreen and the rare twinflower.

The Asian taiga comprises the world's largest forest; in the USSR alone there are nearly ten million square kilometres of boreal forest, much of which consists of larches. These deciduous conifers are

9

dominant over large areas, especially in eastern Siberia, since they can withstand both the extreme winter temperatures of -30°C, the January mean, and the very low rainfall. Siberia contains the broadest part of the world's coniferous girdle, over 2,000 kilometres wide, and it is here, at almost 70°N, within the Arctic Circle, that the Dahurian larch forms stands of the northernmost forests in the world.

The Yenisey River marks an astonishing change from the forests of pine, spruce, Siberian larch and Siberian fir to the west, with Eastern larch and stone pine in the wetter areas, to the forests of silver fir and Dahurian larch to the east. Further south, China's coniferous forests are dominated by Manchu pine, but are also characterised by the presence of such broad-leaved deciduous trees as oaks, maples and elms. The Soviet island of Sakhalin, although lying south of the main belt of taiga, is liable to some seven months of snow per year and has a growing season of only about two-and-a-half months, supporting forests of Dahurian larch and Japanese stone pine. Even on the northern Japanese island of Hokkaido there are boreal stands of spruce and fir.

The temperate climatic zone, with its distinct cold season, has only developed fully in the northern hemisphere; the tip of the South American continent is the only land that falls within the equivalent latitudes south of the equator. As a result conifers are not

the main problem for the species of both habitats being insufficient available water.

The leaves are perhaps most modified, being shaped like needles in the majority of coniferous species. As with the spines of desert cacti, this narrow profile reduces the leaf's surface area in relation to its volume and this serves to reduce water loss in an environment where water is frozen for much of the year. In addition, the leaves possess thick, resinous cuticles, which prevent water loss through the cell walls. The stomata, through which water vapour might also escape, are located in tiny pits lying within a groove in the underside of the leaf; these depressions reduce air movement around the stomata, the still air becomes humid and further diffusion of moisture from the leaf is avoided. In very severe weather, when the soil is frozen and the roots cannot take up water, the stomata can be closed completely.

The leaves of most conifers are evergreen and are not shed each autumn in the manner of deciduous trees. The life span of each needle varies according to the species of conifer and ranges from about three to twelve years; they are discarded a few at a time during the growing season. The advantage of having evergreen leaves is that photosynthesis may commence immediately conditions become favourable, without the tree first having to produce new leaves. Larches, on the other hand, are deciduous and it may be this which

The thick-barked ponderosa pine (**1**) is particularly resistant to the fires which regularly sweep the lower mountains of North America's Pacific coast and is thus a common component of their mixed forests, along with redwoods, sugar maples and beeches.

2

1

widespread in the southern hemisphere, except in a few mountainous regions. There are no true pines, firs, spruces, cedars or cypresses native to the southerly continents, although fragmentary forests are sometimes formed by similar species, such as the yellowwood of South Africa, Tasmania's King William pine and the Australian hoop pine. The Chilean cedar, Patagonian cypress, candelabra tree and the araucaria, more commonly known as the monkey puzzle tree, all grow in the Andes.

Conifers have developed an extraordinary range of physical and chemical adaptations whereby they are able to survive the extreme climatic conditions in such high latitudes. Many of these adaptations are also found, surprisingly enough, in the plants of arid lands,

3

World Distribution of the Great Grey Owl

helps them to survive under even more severe environmental conditions, since the tree in effect becomes dormant. This, together with the fact that larches can produce new leaves very quickly, probably accounts for their extreme northerly range in Eurasia.

Furthermore, the needles of coniferous trees are normally dark green in colour, enabling them to absorb the maximum amount of light from a weak, northern sun. Finally, conifers have a high sap concentration, especially in winter, which acts in a similar manner to antifreeze, allowing conifers to tolerate subzero temperatures without the risk of their fluids freezing, rupturing the cell walls and severely reducing the tree's vigour.

But these leaf modifications also have their disadvantages, the main one being their resistance to decomposition once they have fallen from the tree. The breakdown of pine needles into their component chemicals may take many years, partly because of the small surface area for the microorganisms in the soil to attack, partly because of the needles' thick cuticle and partly because of their chemical composition. As a result a deep layer of dead, but undecomposed, needles builds up on the forest floor, which means that the minerals and nutrients accumulated in the trees' tissues are only returned to the ecosystem very gradually. This problem is further exacerbated by the low levels of microbial activity in such cold conditions.

Nearly all conifers have very shallow root systems, extending in a close network for many metres from the base of the tree. Not only is this an adaptation for growing in shallow substrata, but it also allows the roots to start absorbing water immediately the surface layers of the soil begin to thaw in spring, thus permitting the rapid resumption of photosynthesis. This wide, shallow root system also favours the symbiotic relationship that many conifers have with forest toadstools. These fungi are extremely efficient at breaking down the tough, resinous needles of the conifers, thereby extracting their mineral and nitrogen compounds. The contact between the fungal mycelium and the roots' surface then allows the tree to reabsorb these compounds from the fungus. In return the fungus obtains from the tree

Because of their nocturnal habits and unearthly calls, the world's 135 or so species of owl have long been regarded as birds of ill omen. They occur in all biomes other than the marine realm, and on all continents except Antarctica. The distribution of the great grey owl (**4**,**5**,**6**) coincides almost exactly with that of the boreal forests, although the magnificent great horned owl (**3**), with its distinctive ear-tufts, is confined to the New World. This latter bird will attack animals the size of a large domestic cat, although the slightly smaller great grey owl prefers squirrels and lemmings. The North American pine siskin (**2**), like many boreal forest passerines, takes advantage of the summer feast of abundant insects to rear its young, but winters to the south.

sugars and carbohydrates, which it is unable to photosynthesise for itself, since it does not contain chlorophyll.

It seems that no two species of conifer have identical ecological requirements in terms of temperature, light, nutrients and water, as a result of which competition between species is small. This may account for the high occurrence of stands consisting of only one or two tree species, which often extend over vast areas, until environmental conditions change slightly and a new species becomes dominant. In general, conifers make fewer demands on their environment in terms of their nutrient requirements than do deciduous trees, and they are thus able to thrive on the poor, glaciated soils of the north.

Conifer trees are largely conical in shape, with branches growing spirally from a single trunk, the lower ones longer than the upper. Both the trunk and the branches are usually very flexible and can bend almost to the ground under the snow's weight without breaking. The shape of the needles themselves also deters snow from settling. Like the cuticle of the leaves, the bark is thick and pulpy, both to reduce water loss and to prevent excessive fire damage to the living tissues beneath.

The various ways in which coniferous trees have adapted to their particular environment also enable them to grow in other habitats with similar climatic conditions. For example, mountains are often subject to lower rainfall and higher winds than their surrounding areas and this, together with their steep slopes, which can only support poor, thin soils, makes them ideal environments for the growth of coniferous forests. As altitude increases so temperature decreases, to the extent that for every additional 1,000 metres, the temperature drops a further 6°C. Thus, in the same way that the Arctic tree line determines the northerly limit of tree growth, so there is an upper level beyond which trees cannot survive in the mountains, a level which may vary with prevailing winds and aspect.

Despite the similarity of climatic conditions in both mountains and the high latitudes of the northern hemisphere, there are also some obvious differences between them, such as light intensity and length of daylight. As a result, the species found in the more southern mountain regions are usually different from those of the northern boreal forests. The altitude at which coniferous forests begin to grow must necessarily be higher the nearer to the equator the mountains are situated.

The continental ranges of the Carpathian Mountains

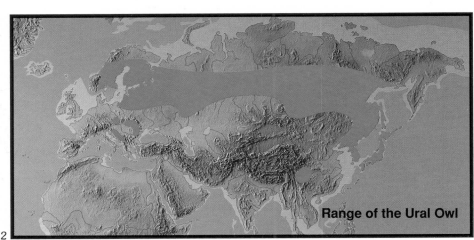

Range of the Ural Owl

The Ural owl (**1,2**), common throughout the Eurasian mixed forests, is a medium-sized bird which feeds primarily on small mammals and hazel hens. Its smallest relative in the region is the pygmy owl (**8**), which measures only about fifteen centimetres from head to tail and, unlike most owls, frequently hunts by day.

in Czechoslovakia are clothed with forests of mountain pine and Norway spruce, whereas in the Pyrenees, which are affected by moist winds from the Atlantic, Scots pines, black pines and firs are more common. In the Alps, spruces are able to grow at heights of up to 1,850 metres, whereas the hardier European larches still persist at levels of over 2,000 metres, as in the Italian Val d'Aosta. Other Eurasian mountain ranges supporting coniferous woodlands include the Apennines of central Italy, the Caucasus Mountains in the southwestern Soviet Union, bordering Turkey and Iran, and the Himalayas. Coniferous forests can also be found on the high mountains of western North America, where they spread southwards from the boreal zone along the Rocky Mountains and the coastal ranges towards Mexico. In the Rockies, eastern hemlock, jack, red and white pines and white cedar are common species, but on the coast, where precipitation is much higher, grand and Douglas firs, western hemlock and Sitka spruce predominate.

Cedar waxwings (**5**), so called because of the colourful waxy secretions produced at the tips of wing and tail feathers, are common summer visitors to the northern forests. Seed-eating crossbills (**4**) remain in the boreal zone all year round, although the search for food has sent birds as far south as the Sahara and Nicaragua.

6

timber for shipbuilding, and indirectly, through the browsing of transhumant sheep and goats.

The ability to survive fire is also far more important in these southern coniferous forests, where lightning is a common phenomenon in summer. Coniferous forests, in which there are large quantities of dry needles on the forest floor containing inflammable resinous compounds, create conditions conducive to fire. The trees themselves actually benefit from regular fires, because the nutrients locked in the dead needles are thereby released into the soil much more quickly. Although deciduous seedlings are usually killed by fire, the buds of the young conifers are protected by tufts of leaves which burn at low temperatures. In addition, some species have developed cones which are tightly sealed with a pitchlike substance and the seeds cannot be released until this has been burnt away. The seeds then germinate in the rich ashes. If regular fires do not

Equipped with short, broad wings and a long tail, the goshawk (**6**) is particularly adept at manoeuvring through the trees in pursuit of avian victims on which to feed its rapidly growing chicks. The bald eagle (**3**) usually nests in conifers near rivers or lakes as fish form a major part of its diet. Unfortunately, the indiscriminate use of agricultural pesticides, which later found their way into the food chain, has dramatically reduced its numbers, making the bald eagle one of America's most endangered species. The male capercaillie (**9**), with its striking blue-black plumage, was long thought to be a different species from the smaller, drab brown female; its clumsy flight suggests that of an airborne turkey.

7

8

9

The ways in which conifers have evolved to guard against water loss and to resist fire also facilitate their survival in warm, Mediterranean climates afflicted by summer drought. In southern Europe and northern Africa the dominant trees are cypresses, junipers and cedars, such as the umbrella-shaped cedar of Lebanon which, despite its name, is today mostly found in the Atlas Mountains of Algeria. In the western Mediterranean, maritime pines are common in coastal areas, growing on sand dunes in areas where the summer drought extends over four months of the year or more. In this species the typically conical shape of the northern conifers is replaced by branches which grow outwards and upwards to absorb the maximum amount of light and, possibly, to direct rainwater towards the roots. In addition, the tree canopy is much more open, permitting a layer of xerophytic herbs, mostly species of grass, to grow beneath. The conifers of this region, however, have been decimated by centuries of human activity, both directly, to provide

10

Coyotes (**7**) are today found from Alaska to Central America. Although originally creatures of the prairies, they have successfully spread into other habitats, including coniferous forests, largely because of their catholic feeding habits. The diminutive stoat (**10**) is a fierce forest predator that will even tackle a hare.

occur, the litter of pine needles will build up and the next fire will be too fierce for even the flame-resistant conifers to survive. Thus the humid, subtropical climates of the southeastern United States support coniferous forests because, despite the higher rainfall, the sandy coastal plain is subject to frequent fires.

For reproduction, the male cones release large quantities of pollen, which drifts through the forest like yellow mist until a chance encounter with a female cone fertilises these unobtrusive 'flowers'. For some species the growing season is too short to allow full maturation of the seeds, which may take a further two summers. When the cones are fully mature, the scales open and release the seeds. Many of these are winged and can drift on the wind for a considerable distance before finally coming to rest some distance from the parent tree, thus avoiding competition with it for the limited water and nutrients. Yet only a minute proportion of conifer seeds survive to develop into young trees. For the most part the nutritious seeds, containing all the energy necessary to produce the first seedling leaves, are consumed by the various animals which have chosen to make their homes in the chilly northern forests.

1

Deer are without doubt the most numerous large mammals of the world's coniferous forests, especially red deer and their relatives, such as the mule deer (7), found only in western North America, and the wapiti (5,6). The dappled coat of a young Eurasian red deer calf (1) blends almost perfectly with the surrounding vegetation. Despite this camouflage almost fifty per cent of calves fall victim to predators within the first few days of life.

2

3

Generally speaking, the animals of the boreal forests are few both in number and variety, limited by the severe winters, low habitat diversity and virtually inedible plant material. Few creatures will eat the tough, acid needles of the conifers, although some members of the bird order *Galliformes*, such as Canadian spruce grouse and capercaillie, have stomachs which can digest them, whilst the small, finch-like grosbeak will nibble at the tender spring shoots. The seeds, however, do provide a viable source of food and are produced in such great quantities that they are available all year round.

Those birds typical of the boreal forests have adopted a variety of different techniques for eating conifer seeds. The crossbills prefer to prise the seeds from the cones before these fall from the tree and they have evolved a beak with crossed mandible tips in order to remove the seeds more efficiently; this feature develops as the birds mature, the young being born with uncrossed beaks. Interestingly, the crossbill has evolved different-sized beaks in populations feeding on the cones of different coniferous species. The parrot crossbills, which are almost always associated with stands of Scots pine, have more robust beaks, since the cones of this species are large and tough. The more slender-billed red or common crossbill is normally associated with spruce and fir forests, whilst the white-winged crossbill shows a marked preference for larch.

The nutcracker, on the other hand, is a sizable bird with a powerful bill that can crush mature cones whole to expose the nutritious seeds within, whilst its close relation, the Siberian jay, although large and heavy, climbs out onto the tips of the branches to feed. Both

these members of the crow family store caches of seeds either in holes in tree trunks or in the snow for consumption during the winter. Thus the specialised seed eaters are able to survive in the coniferous forests all year round, as are small birds such as the crested, willow and Siberian tits and their American counterparts, the black-capped and boreal chickadees.

Mammals also feed on conifer seeds. Red, grey and flying squirrels are able to extract them from the cones whilst these are still on the trees, while ground-dwelling creatures such as American and Siberian chipmunks, wood and bog lemmings, pygmy shrews, Eurasian ground squirrels, spruce and jumping mice, golden opossums and red-backed voles must wait until the mature seeds fall to the forest floor, as must the cumbersome capercaillie and hazel grouse. Many of these animals hoard food during the summer to sustain them during the long winter; the American chipmunk will even store whole cones.

Lemmings breed very quickly in response to favourable conditions. Most conifers produce a bumper seed harvest every three or four years to ensure that some seeds will survive to become young trees. Responding to this abundant food supply, female lemmings may bear up to twelve young per litter and three litters per year; the young of the first litter of the season are also able to breed before the year is out, mating when they are only nineteen days old. However, the following year, when the seed harvest is normal, there will probably be insufficient food to sustain this increased lemming population. Starvation therefore prompts them to undertake long journeys in search of food; many die during attempted river, road or railway

4

The beaver, with its flattened tail, which serves as a rudder, huge incisors, which continue to grow throughout its life, and thick, waterproof fur, is the second largest rodent in the world after the capybara. The two species – the Canadian beaver (3) and the European beaver (2) – are almost identical, classified separately largely on the basis of their geographical distribution.

The American red deer, or wapiti (**5,6**), is today largely confined to the Rocky Mountains, but the moose (**8**), largest of all deer, is found in both European and New World taiga. Both are also, confusingly, called elk.

Tree-dwelling sciurids of the boreal forest, such as the American red squirrel (**4**), and terrestrial species, such as the Uinta ground squirrel (**9**), feed mainly on the seeds of conifers, supplementing their diet with fruit in the autumn.

crossings, which may have given rise to the legend that lemmings commit mass suicide in times of hardship.

Crossbills also travel further afield in search of food in bad winters, characteristically invading beechwoods and gardens all over Europe. But these movements of lemming and crossbill are not considered to be proper examples of migration since they are largely without direction amd are motivated only by hunger. Populations of predators, such as the great grey owl, are also affected, their numbers rising and falling a season behind their prey, the lemmings, just as the lemmings are a season behind the conifers.

The most important of the large herbivores of the boreal forests are the deer, of which there are more species here than in any other type of habitat. Amongst these, the American moose is thought to be the same species as the European elk, although their ranges are widely separated, and the woodland caribou of North America is almost identical to the reindeer of the taiga. Other species include the Altai wapiti, or Asian red

Female brown bears (**2**) are almost always seen together with their cubs, which are born during the winter – blind, toothless, almost hairless and weighing less than 0.5 kilogrammes so as not to exhaust the mother when food is scarce. By the autumn they are ready to hunt with their mother, but do not become sexually mature for another three years.

climb with agility within a few days of birth. The porcupine strips the bark from branches and is also one of the few creatures to feed on young needles. The ability of porcupines to defend themselves against predators by 'firing' their quills probably owes more to myth than reality, although if threatened the porcupine turns its back on its adversary and the quills, normally carried flat against the body, are raised like the hackles of a dog; it may also lash out with its heavily barbed tail.

Because of the scarcity of prey in the boreal forests, carnivores must range over huge areas in order to obtain enough to eat, especially during the winter. The lynx, for example, may travel over an area exceeding 200 square kilometres in search of food. However, many potential victims are able to escape because their hunters will not expend more than a minimum amount of energy in the chase. If, for example, a lynx has not brought down a snowshoe hare within about 200 metres it will probably give up, since the amount of energy to be obtained from eating the hare is outweighed by any further effort needed to catch it. Lynx are able to move effortlessly over the snow due to their flattened, furry feet, which spread its body weight over a greater

Range of the Black Bear

area. Victims such as roe deer, however, often flounder in snowdrifts, becoming easy prey.

The boreal forest predators include the wolf, red fox, racoon-like dog, lynx, wildcat and the now very rare Siberian tiger. Brown and black bears, on the other hand, feed mainly on plant matter, but are not averse to eating meat should the opportunity arise. The grey wolf has the greatest range of any living land mammal other than man, being present in habitats ranging from tundra to desert throughout the northern hemisphere. They prefer large prey, hunting in packs in order to bring down a young or injured moose or caribou trailing behind its normally close-knit herd.

Members of the *Mustelidae*, or weasel family, such as the Siberian and European weasels, the wolverine, American mink, sable, pine marten, American marten and fisher have adapted well to the challenges of the boreal forest. Although similar in body form, each adopts a slightly different method of hunting a variety of prey, thus avoiding competition for the limited food resources. The sable and marten are mainly arboreal creatures, preying on squirrels and small birds, although the sable will also descend to the forest floor to hunt down red-backed voles and may even feed on seeds if prey is scarce. The wolverine, also known appropriately as the glutton, will eat almost anything and is aggressive enough to bring down a young reindeer or elk. The fisher of North America has perfected a technique for

The North American grizzly bear (**1**) is among the largest of the geographical races of brown bear, some males weighing over 600 kilogrammes in their prime. By nature solitary creatures, the males are intolerant of intruders, rivals often coming to blows. Another North American species, the black bear (**3**) differs little from the grizzly; it is smaller, with less powerful feet and claws and a longer muzzle, but coat colour is not a distinguishing feature.

American badgers (**4**) are inhabitants of both forest and prairie. Their numbers are more constant in the former, the human war waged on their main grassland prey, the prairie dog, having reduced the prairie population considerably.

deer, which is very similar to the European red deer, as well as roe deer, musk deer to the east of the Yenisey River and the North American black-tailed and white-tailed Sitka deer.

Elk, or moose, feed largely on aquatic plants in the numerous taiga lakes during the summer, but when these freeze over in winter they take to the shelter of the trees, feeding on bark and scraping the snow from 'feeding craters' on the ground to expose the sparse vegetation below. The moose is the largest of the world's deer, standing about two metres high at the shoulder, although the hindquarters drop away sharply. The males have spreading palmate antlers which may span almost two metres. There are about one million of these majestic deer in the world today, over ten per cent of which live in Sweden.

Another bark eater, albeit one operating some distance above the forest floor, is the North American porcupine, also known as the tree porcupine on account of its arboreal life style; the young are able to

killing tree-climbing porcupines, although it will also kill small rodents and hares, and the versatile American mink can swim, enter burrows and climb trees, enabling it to feed on a wide range of prey species.

The abundance of small, nocturnal mammals may explain why the boreal forests are home to so many species of owl. Some diurnal birds of prey – golden eagles, goshawks and merlins – also favour these undisturbed woodlands. Those owls typical of the Eurasian taiga include the eagle owl, Tengmalm's owl and the Ural owl. In North America the commonest species is the boreal owl, whilst the great grey and hawk owls are found throughout the northern coniferous forest zone. Even such diminutive species as Richardson's owl of North America and the Eurasian pygmy owl are voracious hunters. Others, such as the long-eared and short-eared owls, only inhabit these regions during the summer, preferring to spend the winter where food is more abundant and the climate milder.

The harsh winters of the world's northern coniferous forests create as many problems for the animals as they do for the trees. Size is a distinct advantage in cold conditions as bulk aids heat conservation; it is no accident that the moose is the largest deer, the wolverine the largest member of the weasel family and the capercaillie the largest of all grouse. By contrast, smaller animals have few means of adapting physically to cope with the severe cold and can only survive during the winter by virtue of the thick layer of snow which covers the ground of the northern forests. This layer is formed of a lattice of ice crystals which traps air within and thus insulates the forest floor from extreme temperatures. The heat produced by the slow process of decomposition within the soil gradually melts the lower levels of snow, creating warm tunnels in which small mammals can live and feed quite happily without resorting to hibernation.

Wolves may also take advantage of the insulating

1

2

3

properties of snow by burrowing into drifts to conserve their body heat, but other large animals, such as bears, prefer to sleep through the harsh winter. In order to sustain the body's essential processes during hibernation, sufficient food reserves must be built up during times of plenty, much as squirrels hoard seeds, but *within* the body.

Movement across the winter snows can cause problems for hunter and hunted alike and, to combat this, many animals have broad feet, often with a fringe of hair or feathers around the toes to increase their surface area even further; these include the aptly-named snowshoe hare, the lynx, wolverine and capercaillie. The snowshoe hare, the main prey of many boreal hunters, even changes its coat colour from brown-grey during the summer to white during

the winter, so as to blend in with the snow and thereby escape excessive predation. Moose and woodland caribou are too heavy to travel across the surface of the snow, but can nevertheless move around easily enough on their stilt-like legs, providing the snow remains soft. However, if the top layers thaw during one of the brief warm spells and then refreeze, the crust of ice which forms is hard enough to lacerate the legs of these large deer species; it also makes their excavation of feeding scrapes difficult.

Many birds of boreal forests cannot adapt to survive such bleak conditions. Instead they spend the winter in warmer climes, often as far south as the Mediterranean region or tropical Africa, only migrating north to breed. Such long seasonal journeys are justified by the vast populations of insects which thrive in the northern

The New World tree porcupine (**5**) has a dense protective coat comprising both hairs and spines. Black bear cubs (**1**) learn to climb trees in search of food, especially the contents of birds' nests. Young wild boar (**3**) sport striped coats which provide an effective camouflage to protect them from potential predators.

coniferous forests during the summer. The actual number of species which are able to feed on the tough coniferous foliage is not high, but for those that can, the food supply itself is almost inexhaustible; the larvae of certain insects, which breed in millions, are amongst those able to take advantage of this. There are two particularly common insect species in the pinewoods of the north: the pine beauty moth and the pine sawfly, a member of the wasp family. The larvae of both species consume vast quantities of pine needles and then in turn become food for the influx of breeding birds. In an attempt to escape the eagle eye of their predators, the larvae of the pine beauty moth closely resemble the needles of their host plant.

Many species of migrant birds are brightly coloured and seem rather out of place amid the sombre coniferous trees. In Eurasia they include redwings, fieldfares, willow warblers, redpolls, scarlet grosbeaks, siskins, coal tits, goldcrests, redstarts, Siberian rubythroats, bluethroats and bramblings. This diversity and abundance is mirrored in the forests on the other side of the world, although the species themselves are different, with myrtle warblers, slate-coloured juncos and ruby-crowned kinglets among the more common migrants to the North American boreal forests.

Whilst most of these birds build typical nests, well

softwood lumber industries. These have a considerable effect on the environment, reducing the ecosystem's productivity by depriving it of the essential nutrients stored in the biomass of the trees. Other recent settlements in these northern forests are dependent on mineral exploitation and the lucrative fur trade. The increasing recognition of the value of such wilderness areas for recreational purposes is also likely to have an adverse effect on the wildlife of this remote forest realm.

Coniferous forests represent one of the few habitats still expanding in the world today. Exotic conifers – those not native to the environment in which they are growing – are now widely planted to provide man with timber and with wood pulp for the world's newspapers. But these plantations are a poor imitation of the rich and ancient coniferous woodlands. The trees are normally planted in regimental rows and at such high densities that no light whatever can penetrate to the forest floor, which is consequently barren of life. In Great Britain for example, there are only three native conifers – the Scots pine, juniper and yew – but no less than ten exotic species are being planted for softwoods and pulp, often at the expense of valuable semi-natural habitats such as moor and peatland. This process is being repeated throughout the northern hemisphere.

Fishers (**4**) are arboreal mustelids found only in North America. Despite their name, they prey mainly on hares, squirrels and mice, although an occasional tree porcupine may be taken.

4

5

6

concealed in the branches of the conifers, others seem to find it necessary to have even more security. Woodpeckers, of which the most common representatives to be found in the boreal forests are the three-toed woodpecker in North America and the black and middle-spotted woodpeckers in Europe, painstakingly chip deep nest holes in the trunks of trees. Sometimes, once the woodpeckers have abandoned them, these holes are used by such unlikely residents as goldeneye and goosanders, both members of the *Anatidae*, the family of ducks and geese. Yet other species, such as nightjars and hazel hens, prefer to nest on the forest floor, relying entirely on their camouflage for protection. For this reason, no matter how brightly coloured the males may be, the females are drab and brown, blending in perfectly with the leaf litter.

The isolation and harsh climate of the northern coniferous forests mean that they are amongst the least disturbed wildlife habitats in the world. In many areas they are almost unpopulated by man, probably because of the sheer poverty of the soils, which cannot sustain agriculture, although many stable hunting and gathering communities were able to survive entirely on the products of moose, caribou and fish in past centuries.

Today, boreal forests are the focus of the world's

The wolverine (**2**), also known as the glutton on account of its voracious appetite, is the largest of all mustelids. Despite its clumsy appearance, it is an expert hunter, attacking anything from lemmings to young deer, but also eating carrion, fish and vegetable matter. The Canadian lynx (**6,7**) is largely confined to the North American boreal forest. To cope with the harsh winters, it has a longer, thicker coat and broader feet than species found further south, but still bears the characteristic tufted ears.

7

· CHAPTER 5 ·

MOUNTAINS

Mountains are found on every continent in the world. From the equator to the poles, about thirteen per cent of the earth's land area lies over 2,000 metres above sea level. Lofty and awe-inspiring, snowcapped peaks have challenged man since he first emerged from the African jungle, and many adventurers have lost their lives attempting to conquer these heights. Mountain regions represent some of the last wilderness areas on earth today, standing as 'islands' among the agricultural and urbanised lowlands which form the domain of man. The flora and fauna of the mountains – which survive in a harsh environment that lacks oxygen, has an over-abundance of ultraviolet light and suffers climatic extremes – are among the most beautiful and the least known on our planet.

Left: a New World mountain lion, (top) St Dabeoc's heath and (above) the Picos de Europa in northern Spain.

There is no rigid definition of the word 'mountain', although it usually conjures up two images: firstly, of an area of high land, and secondly – perhaps more importantly – that they rise substantially from the surrounding terrain. The mountains of the world could be defined as land which exceeds 1,000 metres above sea level, but that would be to exclude many of the peaks of the Scottish and Brazilian Highlands – regions which undoubtedly fulfil the usual expectations of the term mountain. On the other hand, many plateau areas of the world exceed this height, including the Great Basin of the western United States, Mongolia's Gobi Desert and most of southern Africa – areas which can hardly be described as mountainous. If the definition is altered to exclude all land below 2,000 metres above sea level, mountains still cover almost twenty million square kilometres of the earth's land surface.

Mountains may occur as isolated peaks, such as Mount Kilimanjaro in East Africa, or the islands of Teide in the Canaries, Mount Apo in the Philippines and Mauna Kea in Hawaii, all of which are the result of volcanic activity. Yet for the most part they form chains or massifs, in which most peaks do not descend to ground level, but instead rise from within the main mountain bulk. Chains such as these are present on every continent. The most impressive is perhaps the Great Divide – the spine of the Americas that extends from the Aleutian Islands to the Scotia Ridge off Tierra del Fuego. In the north, waters pouring from its slopes head for the Arctic Ocean, but along most of its length the Great Divide marks the watershed between the Pacific and Atlantic oceans.

The southernmost half of this chain is the narrow Andean Cordillera, which extends for over 7,500 kilometres from the Caribbean Sea to Cape Horn along the western margin of South America. The Andes contain some of the highest peaks in the world, such as Aconcagua on the Chilean/Argentine border which, at 6,960 metres high, was thought to be the

The scattered distribution of the world's major mountain ranges (2) has been responsible for the isolation of animal and plant communities, and thus a high level of endemism. The alpine marmot (5), for example, is confined to European upland areas, hibernating in burrows during the harsh winter.

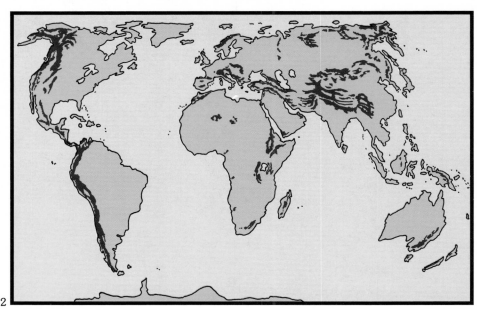

Griffon vultures (4) are among the first to spot a dead or dying animal from the air, descending in large numbers to dismember the carcass with their powerful beaks, whilst the lammergeier (3) – literally 'lamb-vulture' – more often than not makes do with the leftovers.

loftiest mountain on earth until the Himalayas were measured in the seventeenth century. At their northern end, in Venezuela and Colombia, the Andes are composed of three parallel ridges, narrowing into two further south. In the central part of the chain these two lines of peaks are separated at about 4,300 metres by a grassy plateau known as *puna* in Peru and *altiplano* in Bolivia. To the south the Andes gradually taper into a single ridge that decreases in height until, on Tierra del Fuego, the mountains are only about 2,000 metres above sea level.

ridge of the Atlas Mountains in Algeria and Morocco, which border the southern Mediterranean and culminate in the 4,165-metre peak of Toubkal, and the sheer, cloud-covered bulk of the Ruwenzori, or Mountains of the Moon, which straddle the equator on the eastern edge of the Congo Basin. Yet without a doubt the most spectacular African peak is the freestanding volcanic cone of Kilimanjaro, which is also the highest, attaining a phenomenal 5,895 metres.

The Alps are the main European mountain range, spanning the neck of Italy and containing the highest

Weighing up to 700 kilogrammes and bearing formidable metre-long horns, wild yak (6) are rather scarce today, the majority of those encountered in the Himalayas being domesticated creatures, used by the Tibetans for food, clothing and as pack animals.

6

In the northern landmass of the Americas the Great Divide is also represented by an immense chain of mountains along the west coast, but here, although they are much broader than the Andes, the peaks are less lofty. In some places they ascend so gently from the western plains that the Great Divide is all but indistinguishable from the ground. Two main ranges make up the chain, of which the Rocky Mountain Range is the greatest, extending some 5,000 kilometres from the Brooks Range in Alaska down to Mexico's Sierra Madre, and containing no less than sixty lesser ranges. The second main chain comprises the mountains of the Pacific Rim, lying between the Rockies and the sea, and also made up of numerous individual chains, including the Sierra Nevada and the Cascade and Coastal ranges. The eastern side of North America boasts a third range – the Appalachian Mountain Range – which runs more or less parallel with the Atlantic coast between Labrador and Georgia, and includes the Blue Ridge and Great Smoky mountains. North America's highest peak lies at the northern extremity of the Great Divide – Mount McKinley, at 6,194 metres, dominates the Alaska Range

The Ethiopian Highlands, the largest area of mountain massif in Africa, rise to a maximum of 4,260 metres at their northern end in the peak of Ras Dashan. Less extensive chains include the narrow

Distributed across the mountains of Europe and Asia Minor, the graceful and agile chamois (7) lives among the montane forests and meadows during the winter, only ascending to the high alpine pastures during the summer. The North American Rocky Mountain goat (1) is a much more cumbersome beast, the males often weighing up to 150 kilogrammes; their shaggy, white coats are ideal for life above the tree line.

7

peak of the continent – Mont Blanc (4,810 metres). From this central mass radiate other, lesser ranges, such as the Italian Apennines to the south, the Pyrenees and other Iberian mountains to the west and the Carpathians to the east. Further east, however, the land starts to rise dramatically. First come the Caucasus, lying between the Black and Caspian seas. From here onwards there is a continual stretch of land above 2,000 metres as far as the Soviet Pamirs, whose high point is unsurprisingly named Communism Peak (7,495 metres) and to the south of which lie the first ranges of the Himalayas.

The Himalaya-Karakoram chain covers the greatest land area of all the world's mountain ranges. It also contains the highest peak – Mount Everest, which reaches 8,848 metres – and a phenomenal ninety-six of the 109 peaks in the world that exceed 7,300

The ibex (**2**) is an archetypal mountain beast and is found scattered across the mountains of Europe and Asia. The males boast spectacular backward-curving horns up to a metre long, their substantial weight being supported by well-developed forequarters. The musk deer (**4**) of eastern Asia lacks antlers, instead sporting a pair of tusklike upper canines. The males mark their territories with musk from an abdominal gland. This substance is highly prized by the perfume industry. Although moose (**1**) are essentially creatures of the taiga they sometimes visit upland lakes to feed.

metres. These mountains extend eastwards for some 2,700 kilometres from the Hindu Kush in Afghanistan through Pakistan, India, Bhutan and Nepal to China and the Brahmaputra River. Just as the Himalayan chain curves to the south, a further range, the mountains of Kunlun Shan, form a mirror image to the north. Sandwiched in between lies the vast, lens-shaped Tibetan Plateau, with an average height of over 4,500 metres. Drought is a characteristic feature of this barren plain, since it lies in the rain shadow of huge peaks in every direction. Even further north lie the Tien Shan Mountains, which run roughly parallel with the Kunlun Shan. They issue from more or less the same point as the Hindu Kush and Pamir Mountains in the west, but trail away eastwards into the Gobi Desert.

There have been three main mountain-building periods, or orogenies, during the history of the earth.

mountains is that of continental drift, whereby the earth is envisaged to be enclosed in a series of continental plates which 'float' on the semi-molten rock that constitutes the lower layers of the earth's crust. Where these plates impinge upon one another, great stress can build up, especially where two plates are travelling towards one another. The Alps, for example, are thought to have come into being when the African plate, which bore what is now Italy, 'collided', over millions of years, with that carrying the main bulk of Europe, forcing material which had once formed the sea bed upwards in the process. It is thought likely that the Himalayas are the result of a similar impact.

Mountains created in this way took millions of years to form. Volcanic peaks, on the other hand, grow very quickly as a result of the ash and lava flows that emerge from the crater during periods of activity, producing a

The spectacular purple gentian (5) grows in montane meadows and open woodlands in the Alps, Apennines and some northern European ranges. The domain of the ptarmigan (3) lies above the tree line. It is well-camouflaged at all times as its plumage changes from pure white in winter to mottled grey-brown in summer.

4

5

The Caledonian and Hercynian orogenies took place some 450 and 250 million years ago respectively, and were responsible for the formation of the mountains of northern Europe and the British Isles, as well as the Appalachian ranges of North America. During the Alpine orogeny, which started some sixty-five million years ago, all the major ranges of Eurasia came into existence, including the Alps, the Pyrenees and the Carpathians in Europe, and the Caucasus and the mighty Himalayas further east. The North American Rockies and the Andes of South America are estimated to have been formed only marginally earlier, about seventy million years ago, but the Coastal Ranges are estimated to be very much younger, having been in existence for only about fifteen million years.

One theory which attempts to explain the origin of

6

Himalayan griffon vultures (6) are cliff-nesting birds that have been sighted at over 7,500 metres. Unlike many birds, they mate for life, the breeding ritual starting in December, earlier than that of almost any other avian species. Although they may breed in large colonies, a minimum distance of approximately two metres between neighbouring nests is always respected.

distinctive conical shape. Volcanos, of which there are about 600 active examples in the world today, are also the visible result of unseen stresses in the earth's crust, mainly occurring along the edges of the continental plates. The Mid-Atlantic Ridge, for example, is another example of two plates in opposition. Some of the world's highest mountains – if measured from base to peak – are found here, although only the summit may emerge above the surface of the ocean.

As might be expected, by rising above the level of the surrounding land, mountains possess rather different climatic conditions from the plains. Our planet is warmed as air molecules subjected to radiation from the sun become more active, rubbing against each other and creating heat. As the density of air molecules is reduced at high altitudes, they collide less frequently and as a result the air is considerably colder at the top of the mountain than at its base, dropping on average between 5° and 6°C for every 1,000 metres ascended. In the high Himalayas the temperature may be as low as -20°C. Such low temperatures mean that much of the annual precipitation falls as snow, which may be present all year round in the highest peaks, even at the equator. Precipitation varies with aspect; those slopes facing the prevailing winds will receive more rainfall than those on the leeward side, which lie in the rain shadow. Mountains are also scoured by high velocity winds, which may reach 300 kilometres per hour in the Himalayas.

In many of their aspects – low temperature, high winds, prolonged snow-cover – mountain environments resemble those of the northern coniferous forest and tundra habitats, but unlike these northern regions, most mountains, especially those at the equator, are subjected to very high levels of ultraviolet radiation from the sun. This is partly due to the thinner atmosphere at altitude and partly because the sun's rays strike the earth closer to the perpendicular, and thus more powerfully, than at higher latitudes. There

The apollo butterfly (2), with a wingspan of up to eight centimetres, is a spectacular mountain species found in all the major mountain ranges from Spain to central Asia. The larvae feed on montane plants such as stonecrops and houseleeks.

is also a marked difference in day length between poles and equator. At certain times of year a tundra 'day' may consist of twenty-four hours of darkness, at others, twenty-four hours of continuous light, but extremes such as these do not occur at lower latitudes, so most mountains experience greater diurnal rather than seasonal variation in their physical conditions.

Rapid evolution is a common feature of mountain species since extremes of heat and cold, high levels of solar radiation and oxygen deficiency all accelerate the rate of spontaneous mutation, thus providing great potential for genetic change. Since mountains do not form a continuous highland habitat, but instead are almost like islands in a sea of lowland plains, it is not surprising that they are home to large numbers of species which may occur nowhere else: endemic

species, which may even be confined to a single mountain peak. Of a total of 4,000 plant taxa occurring in the Alps, over 200 occur nowhere else (a full five per cent of the flora), whilst the Pyrenees boast about 180 endemic species. In the mountains of Europe, there are many examples of flower species of the same genus, similar in many ways, that are each confined to their own small range; the snowbell *Soldanella minima*, for example, is endemic to the Alps, *S. villosa* is found only in the Pyrenees and *S. carpatica* is restricted to the Carpathians. The same applies to many other species, such as gentians, primroses, harebells and columbines.

Although mountains occur on all continents, in almost all latitudes, and in a great variety of climatic conditions, it is possible to distinguish a basic pattern

In Europe the many high montane grasslands are covered with sheets of trumpet gentians (3) and wild daffodils (4) in spring. Gentians, such as the Nepalese *Gentiana depressa* (5), are typical mountain plants, low-growing and early-flowering to cope with the winter snow cover and short growing season.

5

generally peter out into a sub-alpine, dwarf pine and shrub zone before reaching this altitude. The true alpine zone is encountered between about 2,600 and 2,800 metres, where no trees are able to grow and the vegetation consists largely of herbaceous perennials, especially grasses. In its upper reaches this alpine zone may pass through a transitional sub-snowline belt before giving way to permanent snows at heights of between 2,500 and 3,200 metres, depending on the aspect, maximum height and oceanic or continental location of the mountains.

An example of tropical mountain zonation can be seen in the Ruwenzori Range of East Africa, the highest peak of which is Margherita, at 5,109 metres.

Cushion-forming species are also common in the mountains, the compact shape being designed to raise the temperature of the plant and thus artificially prolong the growing season. Such widespread species as King of the Alps (**7**), a close relative of the forget-me-not, and moss campion (**9**), which grows at heights of up to 3,700 metres, produce such a profusion of flowers that they resemble small, brightly-coloured hedgehogs in the summer. The attractive white crucifer *Teesdaliopsis conferta* (**8**) is found only in the mountains of southwest Europe, especially the Cordillera Cantábrica of northern Spain, where it favours loose, slaty soils.

6

7

8

9

of vegetation. Several zones can be identified and the identification process largely depends upon the zone's altitude, although its aspect plays a part. The ascent of a high mountain resembles, to all intents and purposes, a much-condensed journey to higher latitudes, during which all the habitats lying between that particular mountain and the pole can be experienced.

Within the vegetation of temperate mountains, such as the Alps, for example, no less than seven distinct zones have been distinguished. The lowest of these is the colline or hill zone that lies between about 550 and 1,000 metres and usually consists of oak forest that grades into sub-montane beech woodlands to heights of about 1,700 metres. The mountain, or oreal, belt above this comprises dense spruce forests – these can extend up to some 2,400 metres, but they

The savanna grasslands lie between 1,200 and 1,500 metres, grading into a luxuriant, tropical high forest. At 2,200 metres the upper montane forest begins, dominated by primitive conifers known as podocarps and large stands of mountain bamboo, which can reach a height of fifteen metres. The constant cloud cover of these slopes creates humidity, so epiphytes are common and branches are frequently draped with hanging lichens. At about 3,000 metres the grassy glades and enormous tree-heath thickets of the Afro-alpine region start to appear. This vegetation type, unique to the East African high peaks and the Ethiopian Highlands, is characterised by giant groundsels and torch lobelias. The zone extends upwards into a barren belt of stunted grasses, mosses and lichens, before giving way to permanent snows between 4,200 and

Another typical upland plant is the delicate pink fairy foxglove (**6**), which grows in the mountains of southern Europe, usually on limestone rocks and screes. The Pyrenean saxifrage (**1**) is also a calcicole, found only in these mountains, where it grows at altitudes of up to 2,400 metres.

4,600 metres. As might be expected, on both tropical and temperate mountains the number of plants and animals present decreases with altitude; above the snow line there is little but ice and rock – terrain in which few creatures are able to make a living.

There are two major limits which are encountered on all mountains: the tree line and the snow line. The level above which trees are unable to grow due to the harsh environmental conditions is similar to the Arctic tree line of the northern coniferous forests, only in this case high altitude rather than high latitude is the limiting factor. Generally speaking, trees are able to grow at greater heights close to the equator, on larger mountain masses and in oceanic climates – that is, where the mountains lie in the path of rain-laden winds from the oceans. The tree line may vary locally by 200-300 metres depending on topography, aspect and soil depth. Of course, it has also been lowered artificially for a variety of reasons in areas long occupied by man. Thus the highest timber lines occur at about 4,700 metres in the continental Himalayas, but drop to a mere 600 metres at the northern end of the American Cordilleras and on Tierra del Fuego, only about 1,200 kilometres from Antarctica.

The snow line, which marks the level above which snow lies throughout the year, is dependent upon more or less the same factors as the tree line, with high latitudes resulting in a lowering of the snow line, and vice versa in tropical regions. Some mountains have

1

2

been eroded to such an extent – or else were never sufficiently lofty – that they do not bear a permanent mantle of ice and snow. Nowhere in the British Isles are snowcapped mountains a summer sight, nor in the Appalachian Mountains of North America. Like the tree line, the level of the snow line varies according to the location of the range in question. In the northern ranges, such as the Rockies and the European Alps, permanent snows commence between 2,500 and 3,800 metres. In the more southerly, but less oceanic, Pamir and Tien Shan ranges it lies at around 4,000 metres, while in the temperate Szechwan region, on the other side of the Himalayas, permanent ice cannot

persist below 5,500 metres. The highest snow lines in the world are in the equatorial Andes, starting at around 6,500 metres.

High mountain climate is thus one of extreme temperature variation – bitter cold at night and searingly hot by day. Here there are high levels of ultraviolet radiation and low availability of atmospheric gases. The slopes are scoured by constant and often violent winds, soils are unstable and long winters of heavy snowfalls are often combined with summer drought. As in the polar regions, such environmental factors combine to produce a very short growing season in the high mountains. Many plants, such as purple crocus, pheasant's-eye narcissus, spring gentian and long-spurred pansy, are primed to start their reproductive cycle the moment optimum conditions return, producing flowers and setting seed only days after the snow has melted. Generally speaking, the flower buds are fully formed by the end of the previous season, overwintering in a manner which will allow them to burst into bloom at the first opportunity.

All in all, annual plants that overwinter as seeds are rare at high altitudes, since a brief growing season does not always allow them time to complete their life cycle. In the Rocky Mountains, for example, of over 300 plant species which grow above the tree line, only two are annuals, and in Scotland the only true alpine annual is the snow gentian. The typical plants are perennial and do not rely solely on seed production for the survival of their species. Instead they either

The grassy glades which lie between snow and tree lines on the high volcanic peaks of East Africa are set with giant groundsels (**1**) which may reach six metres in height. The isolated rocky hills, or kopjes, that stud the African plains are patrolled by handsome black and white Verreaux's eagles (**5**) searching for unsuspecting rock hyraxes. The magnificent golden eagle (**3**), one of the most widespread eagles in the world, is typically seen soaring over the mountains of Eurasia. It normally produces two eggs, though if both hatch the first-born usually dispatches its sibling within the first few days.

reproduce vegetatively, sending out runners and rhizomes to produce new, independent plants, or else, like viviparous fescue and bulbous meadow-grass, the flower heads yield tiny new plants in place of seeds. The drooping saxifrage is rarely known to flower, maintaining its population instead by producing axillary bulbils, each of which is able to overwinter like a fat, reddish seed. Thus a perennial habit is a better insurance policy, allowing the plants to wait for a good year to flower and reproduce without endangering their survival.

Despite the incidence of strong winds in the mountains, few flowers, apart from grasses and sedges, use this as a method of pollination – for example, only sixteen per cent of the flora of the Alps is wind-pollinated. The majority of high-altitude herbs have large, brightly-coloured flowers to attract insects, although some species have dispensed with the need for pollination altogether. These apomictic plants, including some species of lady's mantle, are able to produce viable seed without fertilisation, so allowing more time for seed development and avoiding the uncertainty inherent in a reliance upon insect pollinators.

In order to cope with strong winds, most mountain plants adopt a low-growing habit to reduce the drag effect of the blast and shield the plant from damaging

ice crystals and particles of rock carried by the wind. Two main plant types are common among mountain species: hemicryptophytes and chamaephytes. Hemicryptophytes are herbaceous perennial plants with buds at ground level, surrounded and protected by previous years' leaf remains, such as rosette- and tussock-forming species, while chamaephytes are usually dwarf shrubs, mat- and cushion-forming species with their overwintering buds situated less than twenty-

4

5

6

7

The speckled plumage of the female North American white-tailed ptarmigan (**2**) allows her to incubate her eggs unmolested by aerial predators even in the barren uplands of the Rocky Mountains, where cover is scarce. The Andean flamingo (**4**) of soda lakes in the high Andes is easily distinguished by its yellow legs and black-tipped wings. The vivid rose-pink plumage is derived from the consumption of large quantities of the red algae which thrive in the salt-laden waters.

The New World vultures are not very closely related to those of the Old World, although almost identical in appearance and habits. The Andean condor (**7**), one of the largest flying birds in the world, having a wingspan of almost three metres, is found along the whole length of the South American range, particularly in the high plateaux of Peru and Chile. In contrast, the turkey vulture (**6**), the most common and widespread of all American vultures, is more often associated with forests, having a keen sense of smell in order to locate carrion concealed beneath the trees.

weather. Many of the species called 'alpines' by gardeners will succumb to the first severe frost when transplanted to the lowlands! Cushion-forming species, such as purple saxifrage and moss campion, create their own microclimate; temperatures inside the dense mass of vegetative material are more or less constant and may be up to 10°C warmer than the surrounding air. As a result of the heat radiating from within, overlying snow melts more rapidly and the plants are able to flower earlier. The alpine snowbell also utilises this technique; instead of waiting until the snow thaws naturally to produce its flowers, it radiates stored energy as heat and melts the snow itself.

Even where snow persists for much of the year some plants, such as houseleeks and saxifrages, are able to thrive due to their very high chlorophyll content and evergreen habit. If the blanket of snow is thin, light filters through to the plants below and photosynthesis can still take place. A few mountain species, such as glacier crowfoot, are able to photosynthesise more efficiently than their lowland counterparts, building up a higher concentration of sugars in the plant cells. This

Although the raven (**1**) occurs in a wide range of habitats it is particularly associated with mountains. It is the largest songbird in the world, although its distinctive croaking call is hardly melodic. Another corvid, the chough (**8**), with its bright red, downcurved bill and red legs and its acrobatic, buoyant flight, is found throughout the high mountains of Eurasia, nesting in rock fissures in the peaks.

five centimetres above the soil surface. In fact, these low-growing plants are often the height they are as a direct result of their altitude, rather than because they are genetically dwarfed – high light intensities and low temperatures reduce stem elongation and encourage prolific branching, thus producing the characteristic cushion- and carpet-like plants.

An additional advantage associated with a prostrate existence is that during the winter the plant is completely covered by an insulating blanket of snow. In fact, many alpine plants, such as pygmy buttercups and dwarf snowbells, are unable to withstand very cold temperatures, but are not required to do so because they are covered by snow during the most severe

Trumpet and spring gentians (**2,7**) are among the first flowers to appear in the montane grasslands of southwest Europe. Pyrenean fritillaries (**3**) and red pasqueflowers (**5**) emerge later, together with the highly toxic common monkshood (**4**) and the delicate Cantabrian harebell (**6**), endemic to Spain's Cordillera Cantabrica.

9

10

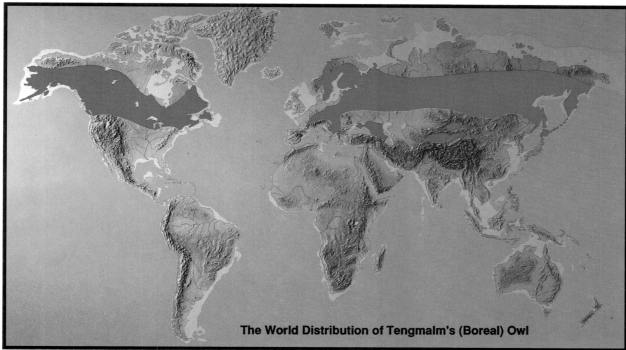

The World Distribution of Tengmalm's (Boreal) Owl

11

Andean geese (**10**), confined to this mountain range, are in fact more closely related to ducks than to true geese. During the summer these snow-white birds nest at between 3,000 and 5,000 metres, but when the harsh winter closes in they gather in large flocks in the lower meadows and pastures. An equally typical bird of high mountains is the alpine accentor (**12**), which prefers rocky open terrain well above the tree line in the European and western Asian ranges. The distribution of Tengmalm's owl (**9,11**) coincides broadly with that of the northern coniferous forest zone, but within this range it favours hilly regions. This small owl can be distinguished from the very similar little owl by its raised eyebrows and the fact that it is nocturnal (except in the Arctic); the little owl's face appears to be set in a permanent frown and it is frequently seen during daylight hours.

12

has a dual function: it acts as a type of antifreeze during cold spells while enabling the root cells to absorb water from the surrounding soil by osmosis in times of drought.

Many alpine plants possess dark pigments in their cells both to protect them from the high levels of ultraviolet light during the summer and to absorb heat from the sun in cold weather. Some species of saxifrage growing on calcareous rocks extract calcium from the soil and exude it as droplets of lime, which encrust the leaves and protect them from excessive insolation. In addition, montane plants are commonly hairy, an adaptation which serves several functions. Firstly, a sleek coating of hairs allows plants to grow at the very

edge of snow-patches without saturating their leaves, since water rolls off them, secondly, hairs reduce the movement of air around the leaf, thus cutting water loss by evaporation, thirdly, this layer of air also insulates the plant during cold weather, and fourthly, silvery hairs reflect the more harmful of the sun's rays. Some examples of particularly hairy plants are the edelweiss of the European Alps and the saussurea of the Himalayas – the latter rather resembles a ball of cotton wool and has been recorded at heights of over 6,000 metres.

Alongside all the obvious adaptations of mountain plants to their environment, there are some species, such as the giant mountain plants of the tropical alpine regions, whose peculiarities as yet defy explanation by the scientists. In an almost perfect example of convergent evolution, tropical African mountains boast torch lobelias and giant groundsels, while the high Andean plateaus have giant espletias – known locally as *frailejones*, or tall friars – and *puyas*. All are members of the *Compositae*, the daisy family, except

Where trees have been cleared to produce grasslands in the European mountains, spring often produces dazzling displays of wildflowers such as early purple orchids and buttercups (**1**). The extremes of temperature and high levels of insolation experienced at altitudes of around 5,500 metres in the Himalayas have endowed the saussurea (**5**) with a dense coat of greyish hairs to deflect ultraviolet light and retain heat.

recorded, the crucifer *Ermania koelzii*, was found at an altitude of 6,300 metres in the mountains of Kashmir.

Like montane plants, animals which inhabit the high peaks display various adaptations to this specialised habitat where food, oxygen and water are scarce, the sun's rays are fierce and temperature variation over a twenty-four hour period may be extreme. The vicuña – a small, South American relative of the camel – has an enlarged heart and lungs and a greatly increased density of red corpuscles in the blood, all of which assist this graceful creature to live in the altiplano and puna

3

4

The slowness with which the Californian condor (**2**) reproduces, exacerbated by damaging human activities, suggests that extinction is imminent for this species. The lammergeier (**4**) is known in Spanish as *quebrantahuesos*, or 'bone-crusher', since it habitually drops animal bones from great heights to expose the marrow. This bird is becoming scarce over the European part of its range, but is faring better in the Himalayas. The lesser, or red, panda (**3**) is a nocturnal, arboreal creature living in the montane bamboo forests of eastern Asia at heights of up to 3,500 metres.

for the puya, which is related to the pineapple. Lobelias may reach eight metres, puyas over nine metres and groundsels up to fifteen metres in height. Although the causes or advantages of such gigantism have yet to be discovered (they may be associated with the high intensity of ultraviolet light, or low oxygen and carbon dioxide levels), these plants also display more typical adaptations to their montane habitat: the enormous flower spikes of the lobelias have long bracts which insulate the flowers from frost, while dead groundsel and espletia leaves remain attached to the stem, forming a warm, protective collar, those of the latter also possessing long, silky hairs.

Above the snow line, when conditions suitable for growth may occur on just a handful of days per year, only such primitive plants as algae, lower fungi and lichens are able to survive. Nevertheless, records exist of flowering plants growing at extraordinary levels: a creeping species of stitchwort – *Stellaria decumbens* – has been reported at a height of almost 6,200 metres in the Himalayas, and the highest angiosperm ever

grasslands of the Andes where, at about 5,500 metres, the atmosphere contains only half as much oxygen as at sea level. The wild yak of the Tibetan plateau also has enormous lungs, since it rarely descends below 4,500 metres, even in winter.

Reptiles and amphibians are generally poorly equipped for a mountain existence, but for the few that have changed sufficiently to survive in such conditions the absence of competition for food and space has made the sacrifice worthwhile. The Andean smooth-throated iguana is able to remain active at temperatures as low as 1.5°C (when most of its reptilian relatives would have succumbed to hypothermia) enabling this lizard to live at heights of up to 5,000 metres in the South American cordillera. The alpine salamander of Europe has avoided the necessity to return to the water – which is often frozen in the high mountains – by bearing live, fully-formed young. The fifteen-centimetre giant toad of Lake Titicaca has gone to the opposite extreme. It has returned to an almost full-time aquatic existence, since although this Bolivian lake lies at about

5

3,850 metres above sea level, its great depth of almost 400 metres means that the water temperature remains more or less at 11°C all year round.

Much like the dark-pigmented alpine plants, some animals (especially invertebrates and reptiles) produce melanistic forms – that is, individuals with such a high level of pigmentation in their cuticle or skin that they are almost black. Such coloration is thought to serve two purposes: firstly, to absorb heat, enabling creatures which are unable to control their internal body temperature to warm themselves in cold climates, thus remaining active, and secondly, to protect the animal against high levels of ultraviolet light. The servals of the African savannas possess black-spotted golden pelts, but in some parts of the equatorial highlands melanistic forms occur. In most cases such animals are not considered to belong to a separate species, although if they become better equipped to survive at high altitudes than those with normal coloration, natural selection would favour black individuals. This form would thus become dominant in mountain regions, possibly even

evolving into a distinct species – which is probably how creatures such as the alpine salamander (which is coal-black all over) came into existence.

Many large mountain herbivores are extremely sure-footed. Chamois, native to the Pyrenees, Alps and Apennines, can reach ledges as much as four metres above them in a single leap and traverse ravines over six metres wide, whilst the African klipspringer can land on a ledge or pinnacle the size of a dinner plate. The bulky yak is less agile than its more graceful relatives, but is also surprisingly sure-footed, whilst the cloven hooves of American bighorn sheep actually separate on landing to provide a better foothold on slippery rock.

Most of the other adaptations displayed by high mountain creatures are designed to fend off the intense cold. The most common feature of vertebrates is a thick insulating coat of fur or feathers; indeed, some are so heavily clothed that the true shape of the animal is almost indistinguishable. The Andean chinchillas, for example, have the densest, silkiest fur known to man, while the wool of the vicuña has such excellent insulating properties that these animals would expire from overheating after running were it not for small bare patches of skin on the inner thigh which allow rapid heat loss if bared to the wind. The yak also has a dense insulating coat, especially luxuriant on its limbs and tail, where heat loss is greatest, and each hair has

The peregrine falcon (1) is one of nature's supreme killing machines, reaching speeds of over 300 kilometres per hour when diving on its victims, usually other birds. The short-toed eagle (4), related more closely to kites than eagles, preys mainly on snakes.

5

altitudes of between 2,000 and 3,500 metres.

The possession of such valuable fur is one that has cost many mountain animals dear, resulting in the near extinction of several species due to excessive hunting. The chinchilla is now confined to a few isolated colonies in Peru following its massacre by opportunistic European trappers in the eighteenth and nineteenth centuries, and another South American mountain-dweller, the vicuña, decreased in numbers from 400,000 to less than 10,000 individuals during the Fifties and Sixties due to similar persecution. Their once wide range is now restricted to a few small herds in Chile, Argentina and Bolivia, but thanks to the creation of a reserve at Pampa Galeras in 1966, there are now over 60,000 of these elegant creatures in Peru.

Dotterels (7) breed in montane and tundra habitats, wintering along the coasts of Africa and southwest Asia. Rock hyraxes (9), denizens of the African kopjes, are the closest living relatives of the elephant. Limestone crevices collect just enough soil for stonecrops (6) to grow.

6

7

8

9

Cougars (5,8), also known as mountain lions or pumas, are equally at home in the desert or above the snow line, feeding on anything from small rodents to deer. The giant panda (3), which feeds mainly on a few species of mountain bamboo, is confined to the mountains of southwest China today; optimistic estimates put the population at around 1,000 individuals. All bamboos of the same species flower and die at the same time, and many pandas starve to death before the new thickets form. Less than 1,000 mountain gorillas (2) survive in the mist-forests of the Virunga Mountains along the Rwanda-Zaire-Uganda borders.

a waxy coating to further guard against snow and rain.

Many other animals show an adaptation to the severe conditions of their lofty habitat. Confined to the northern end of the North American cordillera, the mountain goat has a thick, white coat extending to its knees – an absolute necessity for life between the tree and snow lines. The snow monkeys, or Japanese macaques, are the most northerly monkeys in the world, their thick, shaggy coats equipping them for survival in the highest mountains of Japan. The Andean tapir, the tracks of which have been seen above the snow line, has a soft, woolly coat, particularly thick on its abdomen, but although well-defended against the cold it prefers to live among the shrubby vegetation at

In order to retain as much heat as possible, most mountain mammals have small ears, short limbs and tend to have a spherical body. Where the same species occupies both montane and lowland habitats, the mountain individuals are always much larger, in order to conserve heat more efficiently. The brown bears of the Pamirs, for example, are among the largest in the world. As in the northern coniferous forests, many mountain animals are the largest representatives of their kind; the nayan, or great Tibetan sheep, and the closely-related Marco Polo sheep of the Himalayas, for example, are the largest members of this family, the markhor is the largest wild goat, and the takin is the largest of the goat antelopes. The pattern is the same among the birds of the mountains; only the albatross has a greater wingspan than the huge Andean condor, and the only birds which weigh more than the twelve-kilogram males of this species are bustards, swans and pelicans.

During the winter many montane species hibernate to escape the worst of the weather, but others migrate downwards to the warmer lowlands in order to find sufficient food. The Himalayan black bear spends the summer near the tree line at altitudes between 3,000 and 3,600 metres, descending to the valleys during the winter, but the brown bear of these same mountains prefers to hibernate. Vicuña also display a seasonal migration up and down the mountains (although they never descend lower than about 3,000 metres), feeding just below the snow line in summer and wintering on the arid plateau. The alpine marmot prefers to sleep through the winter, but long-eared pikas are active all year round at considerable heights, storing caches of food in rock crevices to see them through the worst weather. The ptarmigan is one of the few birds which remain in the mountains for the winter; in severe blizzards it will dig itself into a drift and become torpid until the storm has passed.

Several of the animals that remain in the high mountains all year round, like those of polar regions, are able to change the colour of their fur or feathers to blend in with their environment. The snow vole and the

currents are used by birds of prey – bald eagles, turkey vultures, ospreys and smaller species such as red-tailed and broad-winged hawks – during their seasonal migrations across the continent.

The only predatory mammal which lives exclusively in the high mountains is the snow leopard, or ounce, of the Himalayas and other central Asian ranges. It is one of the most endangered members of the cat family in the world, having been hunted relentlessly as a killer of domestic livestock, for its glorious pelt and for sport. It is an animal truly adapted to its environment,

1

2

ptarmigan, for example, grow glossy white coats in late autumn so that they become less visible against the winter snows, thus avoiding predators more easily. The carnivorous ermine also performs this trick; camouflaged in this way, it is able to ambush its victims without difficulty.

Just as some montane creatures migrate to lower levels in the winter, so animals more typically associated with the lowlands ascend into the peaks during the summer, especially small birds, which take advantage of the swarms of insects that appear there at flowering time. In the African equatorial mountains it is not unusual to see savanna herbivores, such as eland and black buffalo, grazing on the alpine pastures. A migration of a different kind takes place along the length of the North American cordilleras, where rising thermal air

3

The Himalayan tahr (**1**) is a large, goatlike creature with thick, backward-curving horns which frequents the montane pastures and forests of this range, especially in Nepal. The males, some of which weigh in excess of 100 kilogrammes, are solitary by nature, only joining the females in the spring to mate. The intermittent distribution of the ibex (**2,3**) across the high mountains of Europe has led to the evolution of separate, but very similar, species, such as the Alpine, Caucasian and common, or Pyrenean, ibex.

predators such as the marbled cat, Temminck's cat and the leopard cat. The true leopard ascends the lower slopes of East Africa's monumental peaks. In Europe there are no carnivores living exclusively on the mountains; only such species as the wolf, lynx and bear haunt the upper slopes, forced into remote mountain bastions by man's activities in the lowlands in much the same way as the puma has been.

Scavengers and carrion-eaters take the place of the actively predatory birds in high mountains. The vultures of the Old and New Worlds originate from unrelated

The high cordilleras which run along the Pacific coast of North America are home to sturdy Rocky Mountain, or bighorn, sheep (**4,5,7**). Both sexes sport curled horns, although these are much more developed in the male and are used to establish dominance within the herd during ritual combat. Despite the fact that a full-grown bighorn may weigh 135 kilogrammes, these beasts are remarkably agile and can be seen leaping surefootedly in the most rugged terrain. The klipspringers, or African chamois (**6**), of the East African kopjes are also extremely acrobatic creatures, though their tiny, cylindrical hooves would not appear to offer much purchase on these rocky outcrops.

4

5

6

7

possessing a thick, pale silver coat with dark rosettes, furry feet – which enable it to travel in snow more easily – and a long tail for balance when traversing steep slopes or springing on its victims. Its prey species, which include blue sheep, ibex and marmots, are fairly few and far between in the barren peaks, so a snow leopard's territory may exceed a hundred square kilometres.

The mountain lion – also known as puma, cougar, panther or painter in various parts of the world – is another high-altitude predator. This animal is not confined to the mountains, but has been driven into them from its natural grassland and forest habitats during a long conflict with man. Nevertheless, it is very agile, being able to leap up to five metres in pursuit of its prey, mostly deer, rodents and rabbits. There are many subspecies to be found in both North and South America, but sadly these are becoming less widespread as a result of their continued persecution.

Other montane carnivores not confined to this habitat include Pallas' cat from the uplands of Tibet, the solitary and nocturnal clouded leopard of Southeast Asia, Tibetan and Canadian lynxes and smaller

8

The solitary, crepuscular Indian muntjak (**8**) is also known as the barking deer because of its harsh warning calls. In the Himalayas these small deer, which measure only fifty centimetres in height at the shoulder, are frequently seen at altitudes of over 3,000 metres. Like the musk deer, muntjak have tusklike canines projecting from the upper jaw, but they also possess short, two-pronged antlers which sprout from two long, horny processes between the ears.

stock and yet have come to resemble each other both in appearance and habit through a process of convergent evolution. New World species include turkey vultures, North American black vultures and the Californian and Andean condors, whilst typical Old World mountain dwellers are griffon vultures and the magnificent lammergeier.

In the absence of trees, many birds of the high mountains are ground-nesting, such as ptarmigans, dotterels and snow buntings in the northern hemisphere and the ornate tinamous south of the equator. A type of woodpecker called the Andean flicker uses its excavating abilities to tunnel into stream banks and soft cliffs in much the same manner as the European kingfisher. Other species, especially vultures and members of the crow family, utilise cliff ledges. In the

1

2

3

Peruvian puna hummingbirds, sierra finches and small doves nest among the leaves of the giant puyas, the vicious spines of these bromeliads serving to protect eggs and young from would-be predators. However, such security may be bought at a high price by unwary nestlings and their parents, as they too can be injured by the spines.

As well as using the plants as a safe breeding place, many Andean hummingbirds feed on their nectar, at the same time performing a much-needed role in the fertilisation of plants such as the giant espletias and puyas, since insect pollinators are scarce. The sunbirds of Africa play the same part as hummingbirds in the life of mountain plants, although they prefer to perch rather than hover whilst feeding on the giant groundsels and lobelias, and many supplement their diet with insects.

During one of the Everest expeditions an alpine chough was seen taking off from a ledge over 8,300 metres above sea level, and both lammergeiers and griffon vultures have been recorded at heights of over 7,500 metres – again, in the Himalayas. Here too the bharal, or blue sheep, regularly frequents slopes above 5,500 metres, and the chiru, or Tibetan antelope, and the wild yak are among the highest-living mammals in the world – denizens of the world of permanent snows above 6,000 metres. Puma tracks in the New World have been recorded at heights of over 5,600 metres,

4

whilst the snow leopard is known to search for prey well above the snow line at some 6,000 metres during the summer. Among the invertebrates, several species of spider-like harvestmen are known to live at over 5,500 metres, using their long legs to raise their bodies above the ice. Indeed, some species of insect are so well adapted to life above the snow line that they cannot survive at temperatures above freezing point.

Many of the world's larger mountain creatures are so well adapted to life at high altitudes that they are unable to exist anywhere else, and because of the natural scarcity of food resources, most are present on the mountains only in low numbers. Today, these inhabitants of the high lands are among the most

Below the Afro-alpine grasslands, which support giant groundsels (**1**), lie dense montane forests, the haunt of the serval (**4**). The favourite prey of this slender cat is the subterranean mole-rat, whose burrows are located with the aid of the serval's extraordinarily large ears.

Wet meadows in southern European mountains often support stands of the highly poisonous white false helleborine (**5**), whilst the sixteen-centimetre-long black alpine salamander (**3**) is found in the Alps and mountains of Yugoslavia and Albania at heights of up to 3,000 metres.

endangered species on earth, largely because of man's activities, both agricultural and recreational. The mountain gorilla, the world's largest primate, occurs only in the montane rainforests and Afro-alpine grasslands of the volcanic Virunga Mountains on the borders of Uganda, Zaire and Rwanda in Africa. They commonly live at altitudes of 3,000 to 4,000 metres, but often forage at much greater heights. Due to the destruction of their forest environment and the activities of poachers, the mountain gorilla has been brought to the brink of extinction; in 1971 there were less than 1,000 individuals in the wild – a number which has undoubtedly decreased since this time, despite the efforts of Dian Fossey, an American scientist who dedicated her life to understanding and protecting these magnificent beasts.

5

6

7

The Himalayan black bear (**6**) is unusually aggressive for a member of the *Ursidae*, with a much better developed hunting instinct than many of its kind, and thus includes a greater proportion of flesh in its diet. During the summer it lives in forested habitats at altitudes of around 4,000 metres, but descends to the foothills of the Himalayas during the winter. The North American pika (**2,7**) is also known as the calling hare because of its peculiar ventriloqual cry. Although it superficially resembles a rodent it is in fact more closely related to rabbits and hares.

Some of the world's most endangered large mammals may also be found in the Himalayas. The Tibetan gazelle or goa may well become extinct in the wild as only about 150 remain in Ladakh, in Kashmir, and there are only about 350 hanguls or Kashmir stags left in the middle-altitude forests of Dachigam. Less than 2,000 wild musk deer survive today, having been decimated by the demands of the perfume industry (for the musk pod which the male bears under the abdominal skin) and also for their curious, downward-projecting upper canines, which are popular as jewellery. The last truly wild yaks are found only in the Changchenmo Valley in north Ladakh and in a few passes in eastern Kumaon, while the Marco Polo sheep, much prized by

Edelweiss (4) is the epitome of a mountain plant, protected from the elements by its woolly, grey leaves. Rhododendrons (3) may appear less suited to the rigours of life at the top, but many species adorn the Eurasian peaks.

trophy hunters for its enormous, spiralled horns, is today one of the rarest of all wild sheep – only a few survive in remote regions of Ladakh. The markhor, a wild goat bearing magnificent, spiralling horns up to a metre in length, is named from the Persian for 'snake eater', since legend has it that it particularly relishes snake meat. Today, they are found only in the northwestern Himalayas, and the mountains of Kashmir and northern Pakistan.

Other rare and threatened large herbivores include the mountain zebra in southern Africa, of which there are two subspecies: the Cape Mountain zebras number only about 200 individuals today, all of which are in reserves in southern Cape Province, and the larger Hartmann's zebra, of which there are about 7,000 remaining in the mountains of Namibia and southern Angola. Further north, the last few mountain nyala frequent the sub-alpine zone of the Ethiopian Highlands. This graceful antelope was only discovered at the beginning of the century and was thought to have become extinct several decades later; it is now estimated that 4-5,000 individuals survive among the most remote peaks.

5

The mountain zebras (2) of southwest Africa are much smaller than their plains-dwelling counterparts, and considerably more agile. The mule deer (7) is gradually expanding out from its original chaparral habitat on North America's Pacific coast and into the conifer forests of the surrounding mountains. The arboreal crested rat (1), which lives in the montane forests of Kenya, possesses a dorsal strip of long black and white hairs which is erected in times of danger. Many separate races of grey wolf have been identified across its vast range; it is thought that only 500 to 600 individuals of the Iberian subspecies (5) remain.

6

7

The Californian condor is undoubtedly the rarest mountain bird in existence today; in 1980 about thirty individuals were left, only a third of which were of breeding age. Reduced by hunting, air traffic and pesticides, and having an exceptionally slow breeding rate anyway, this bird has now reached such low numbers in the wild that it is doubtful whether it will survive, despite vast financial resources ($25 million over forty years) dedicated to its protection and recovery.

Mountains are highly valued as regions where mankind can escape from the pressures of modern life and absorb the essential spirit of the wilderness. It is perhaps for this reason that a large proportion of the world's national parks have been created in mountain areas. Nevertheless, the spread of man across the earth threatens wildlife in many parts of the world, and montane habitats are no exception. Due to the restricted distributions of many montane animals and plants, whole species can be wiped out by such insensitive processes as forest destruction for fuelwood, as in the mountains of Nepal, or the development of a new ski resort in Europe. Already mountains represent the last refuge for many animals which have come into conflict with man, especially the predatory beasts. As these pressures become more intense, the animals will have to retreat even higher into the peaks, sandwiched between man and an inhospitable land of permanent snow, until eventually they will not be able to tolerate the harsh climatic conditions any longer and will disappear from our world for ever.

8

The Rocky Mountain goat (6) ranges from the snow line to sea level during the course of the year, often descending to the coast to obtain salt, but the yak (8) rarely strays from its Himalayan mountain habitat, often grazing at altitudes of over 6,000 metres.

· CHAPTER 6 ·
ISLANDS

Islands not only support some of the richest assemblages of animals and plants in the world, but they are also home to some of the most bizarre creatures to have ever existed on this planet. Island denizens range from such living fossils as the tuatara lizard of New Zealand, to the last survivors of the most primitive of all mammals: Australia's duck-billed platypus and spiny anteater. The isolation of remote islands has enabled creatures long-extinct on the mainland – the lemurs and tenrecs of Madagascar and the shrew-like solenodons of the Caribbean islands, for example – to survive in relative tranquillity. But discovery and inhabitation of the world's islands by man, and subsequent environmental pressures exerted by introduced domestic livestock and habitat destruction have resulted in the extinction of much of their indigenous wildlife. Giant flightless birds, such as the moa of New Zealand and the dodo of Mauritius, have long since ceased to exist, and many other animals and plants, often confined to a single island, are in danger of vanishing for ever.

Left: rockhopper penguins, (top) elegant and royal terns and (above) a grey seal.

An island can be defined very simply as a piece of land permanently surrounded by water. There are estimated to be about half a million islands in the world today, although this figure is by no means constant. New islands sporadically emerge from beneath the waves and old ones are continually eroded by wind and wave into nonexistence. Broadly speaking there are two types of islands: those which have split off from a larger landmass – continental islands – and those which rise directly from the sea bed – oceanic islands.

It is hypothesised that the earth's continents were once aggregated into a single supercontinent known as Pangaea. About 180 million years ago this enormous landmass began to fragment, and the continents gradually 'drifted' into their present positions. At this time, when birds, amphibians and reptiles were widespread but mammals were still comparatively local in their distribution, it is thought that many of the world's continental islands were formed. The land distribution is thought to have been a disintegrating movement outwards from Pangaea – like slow-motion shrapnel radiating as from an explosion. Continental islands may also arise following the collapse of the middle section of a peninsula due to relentless wave action. An example of this is the separation of Trinidad from Venezuela. The flooding of low-lying land as a result of a rise in sea level may also create islands, as is the case with the Balearic Islands, which represent the unsubmerged peaks of the southern Spanish mountain range which once extended across the Mediterranean basin. All such continental islands inherit a 'living cargo' of animals and plants when they are separated from the mainland.

1

2

Oceanic islands, however, rise directly from the sea floor and have never been joined to a continent. Most are volcanic in origin – such as the Hawaiian chain in the Pacific, and the Lesser Antilles of the Caribbean – and lie along lines of weakness in the earth's crust: on the edges of the continental plates. Where these plates are slowly moving towards, away from or past each other, great stress builds up, often released in the form of volcanic activity or earthquakes. The Mid-Atlantic Ridge is an example of one such zone, stretching from Iceland to Tristan da Cunha. More spectacular still is the 'ring of fire', which encircles the Pacific Ocean. There belts of volcanic activity are responsible for, among others, the island chains of the Aleutians, the Philippines, Micronesia and Polynesia.

Volcanic islands rise from the sea, pristine and sterile, to be gradually colonised by life from other parts of the world. Coral reefs may start to develop on the flanks of tropical volcanoes soon after their formation. In many cases such islands suffer from gradual subsidence due to the unstable nature of the sea bed in volcanic zones, eventually leaving behind them only coral atolls as testimony to their former existence. Perched on a flat-topped submarine volcano, the 40,500 hectares of small islands and coral atolls that make up Pearl and Hermes Reef, for example, are all that remains of the Hawaiian archipelago's oldest island.

3

The main factor to affect the subsequent arrival of fauna and flora on oceanic islands is the proximity of other landmasses and the richness of the animal and plant life of these sources. But it is by no means an instant process. On the most remote islands centuries may pass before enough plant species become established to allow terrestrial animals to thrive, and the influx of such animals themselves can spread over millennia. The first colonisers must be hardy organisms, capable of survival in the harsh conditions which are the inheritance of such islands; barren lava and ash form a steep-sided cone which is immediately exposed to erosion by wind and sea.

Penguins, such as the rockhopper (7), were once considered to be archetypal Antarctic birds, but the discovery of the Galápagos penguin (6), confined to an equatorial archipelago, led to the realisation that penguins are adapted primarily to an aquatic lifestyle and only incidentally to polar conditions.

Ice plants of the genus *Mesembryanthemum* (4) are largely native to the Cape region of South Africa, but have been widely naturalised on islands such as the Canaries at the expense of the local flora.

Surtsey, named after the Norse fire giant Surtur, was the first island to be born in the North Atlantic for over 1,000 years. It first appeared above the surface of the sea in November 1963 about forty kilometres off the coast of Iceland, and sporadic eruptions continued for a further three years. Only six months after the start of the eruption, bacteria, fungi, the seeds of several shore plants and a species of fly were all discovered on the island. Moreover, seabirds and seals were making use of its shores as a resting point. By 1973, just ten years after its birth, thirteen species of vascular plant and some sixty-five types of moss had successfully established themselves there, and 158 different types of arthropod had been observed, although only a few became permanent denizens.

4

5

6

7

Island faunas often contain a high percentage of endemic creatures which have evolved independent of their mainland ancestors. The Malayan bullfrog (2) occurs only in the relative isolation of the Malayan peninsula, and the diminutive golden frog (5) is one of a number of amphibians unique to Madagascar. Yellow warblers (3) are typical insectivorous birds of the Galápagos Islands, while the elegant swallow-tailed gull (1) breeds in large numbers on the same archipelago, deriving its piscine food from the rich surface waters of the cool Humboldt current.

Cracks in the solidified lava of young volcanic islands collect windblown debris, are colonised by small plants and eventually create a suitable habitat for species such as Hawaiian palms (**1**). The néné (**7**) is Hawaii's largest endemic land bird; since its recovery from the brink of extinction, it is also the state emblem. It probably evolved from the Canada goose, although today it is better adapted to walking than flying or swimming and lives in sparsely vegetated volcanic uplands. The prehensile-tailed skink (**5**) is an inhabitant of the Solomon Islands.

1

2

3

But Surtsey surfaced in a region of the world which is relatively poor in animal and plant life, since much of the land is permanently covered with ice. Tropical climates, on the other hand, support a greater number of species and are thus much more conducive to successful island colonisation. Krakatau, lying between Sumatra and Java in Indonesia, literally blew apart in 1883. During the three-month eruption, nearly forty cubic kilometres of rock were blasted away and the resultant ash clouds caused spectacular sunsets all over the world. The centre of Krakatau was annihilated, but three small islets remained on its periphery. All life had been scoured from their slopes in the violence of the eruption, and yet, after only three years, eleven species of fern and fifteen vascular plants had recolonised Rakata – one of the islets – while after fourteen years, no less than sixty-one species of plant and 132 different insects and birds were present. By 1933, 271 species of plant, thirty-six birds, five lizards, three bats, a rat, a python and a crocodile had become established.

Animals that are adapted to life at sea are among the first island colonisers. Seals, marine birds and turtles are frequently encountered on even the youngest islands. But the ocean currents are also responsible for bringing terrestrial life to such shores. The coconut palm is possibly the best example of a plant whose seeds are designed for dispersal by sea. Coconut trees are found fringing beaches above the strandline

4

throughout the tropics, often leaning seaward in order to shed their fruit within reach of the next tide. Each coconut is enclosed in a thick, fibrous coat which traps air and can withstand up to four months drifting with the ocean currents without becoming waterlogged or spoilt by the salt water. When the coconuts near land they are washed up by the tide in exactly the right habitat in which to sprout, develop into trees and start the cycle all over again.

Since most terrestrial and freshwater organisms are killed by prolonged exposure to salt water, one would think it impossible that they should ever be able to colonise remote islands. But scientists, by trailing nets from aeroplanes, have established that at around 3,000 metres above the surface of the earth there exists a sort of 'aerial plankton' made up of tiny animals, their eggs and the minute seeds of many plants. Ferns, for example, have minute spores, and many microscopic orchid seeds are surrounded by air-filled cells to make them more buoyant. Other plants, especially members of the daisy family, utilise 'parachutes' of silky hairs to catch the wind, whilst some tree species, such as maples and dipterocarps, have broad 'wings' surrounding the central seed to provide lift.

But island colonisation is probably influenced not so much by these permanent members of the air

other species, such as land snails or plants with barbed seeds, to become established. These are transported from the mainland or other islands either in the mud which adheres to the birds' feet, or trapped in their feathers. Some tough seeds are even able to germinate after passing through the digestive system of birds, and are accordingly transported to new habitats in the stomachs of their predators.

But getting to an island is only the beginning of the story; finding a suitable habitat in which to thrive, or locating a mate, is by far the hardest part. Cacti are often pioneer plants on lava flows since they have light seeds which are able to germinate in even the driest conditions with little or no soil. The opportunistic species which are normally found in transient habitats – those that suddenly flourish when a fallen tree creates a gap in the forest canopy, or can rapidly colonise newly emerged mudflats after a river changes its course – are primed for instant reproduction, and can increase their population rapidly when presented with suitable conditions. They have a better chance of survival once they arrive on an island, as do animals that eat a wide range of foodstuffs. The most successful bird colonisers are species that normally congregate in flocks, since there is a better chance of a breeding pair being transported to an island together. Solitary birds, such as woodpeckers, are usually absent from remote islands

White terns (**2**), denizens of many of the world's tropical islands, are more usually called fairy terns, though this name strictly belongs to another species. South American terns (**4**), on the other hand, breed in huge colonies on the subantarctic Falkland Islands.

5

6

7

8

Marine mammals frequently make use of small islands for resting and breeding since they are often free of large terrestrial predators. Southern elephant seals (**3**) are found on the shores of the Falkland Islands, as are South American sea lions (**6**). Fur seals are also common on islands around the world, particularly in the southern hemisphere; the Galápagos fur seal (**8**) is the smallest of all southern fur seals. Seabirds such as king cormorants (**9**) also take advantage of the isolation of islands to rear their young in safety.

streams as by sporadic cyclonic storms. While violent winds transport small birds, insects and other creatures to remote islands, the seas are also playing their part. Huge rafts of vegetation – shattered tree limbs, uprooted shrubs and other debris – are swept away from shores of hurricane-stricken lands and may travel hundreds of kilometres across the oceans before they finally disintegrate. Sometimes they may encounter land before breaking up, depositing a whole host of new animals and plants on a virgin shore. In this manner terrestrial invertebrates and reptiles are able to colonise oceanic islands. However, amphibians, due to their greater intolerance of salt water, are rarely present on the more remote islands of the world.

The utilisation of even the newest islands by migratory birds as resting points inadvertently allows

9

since the likelihood of both a male and a female arriving on the island within the life span of one individual bird is almost nonexistent.

Seabirds are among the first and most successful colonisers of remote oceanic islands since they are not dependent upon the presence of vegetation for food, deriving all their nourishment from the ocean. The general absence of predators enables such birds to take advantage of even the tiniest atoll; breeding colonies often comprise millions of birds. For instance, red-footed boobies and frigatebirds breed in the low mangrove swamps at the eastern end of Aldabra, a remote outpost of the Seychelles in the Indian Ocean. The tiny coral island of Aves, in the Caribbean, houses three quarters of a million noddies and sooty terns, despite the fact that it is little more than a glorified beach. Pearl and Hermes Reef is one of the principal breeding grounds of the endangered Hawaiian monk seal, as well as of large numbers of green turtles. Additionally, tens of thousands of seabirds, including tropicbirds, wedge-tailed shearwaters, sooty and white terns, noddies, and blue-faced (masked) and red-footed boobies, also nest and rear their young there.

There are thus two main ways in which island biota become established: either they are already present, courtesy of the mainland flora and fauna, or they arrive independently by means of various different dispersal mechanisms. Continental islands can, of course, be populated by both means, but oceanic islands are limited to chance colonisation from afar. However, once present on an island or archipelago, the animals and plants form part of an isolated community that is no longer subject to the moderating genetic influences of the mainland populations. And so they begin to change.

The term adaptive radiation has been coined for the process whereby a species, having successfully colonised an island, will spread out over the years to take advantage of the various habitats present, usually changing physically in the process the better to exploit

each one. The best known example is that of Darwin's finches on the Galápagos Islands. The original birds are thought to have been blown across from South America during a freak storm at a time when the islands had already been colonised by plants and invertebrates. The archipelago, which probably came into existence about three million years ago, comprises sixteen major islands and numerous offshore islets, some of which are still active volcanoes .

Tremendous variation in environmental conditions prevails among the Galápagos Islands, ranging from virtual desert to humid mountain valleys. It is likely that no other birds were present when the finches arrived, so they were able to extend their range throughout the entire archipelago. From the original seed-eating stock,

King penguins (2) live on the ice-free subantarctic fringes, the largest breeding colonies being on South Georgia, Kerguelen, Macquarie and Marion. There are only about 1,400 Hawaiian monk seals (3) in existence today following their brush with extinction in the 1800s.

no less than fourteen species can be distinguished today, each with a different manner of feeding and each generally occupying separate islands. They differ mainly in the shape of their beaks. Some, like the warbler finch, have evolved tweezer-like beaks to pick insects off leaves and twigs. Others, such as the large

Darwin's observations of the adaptive radiation of finches on the Galápagos Islands provided him with the basis of his theory of natural selection. Among the more bizarre feeding habits of these finches are the woodpecker finch's (**4**) use of cactus spines or twigs to extract invertebrates from cracks in trees and the vampire finch's (**5**) bloodsucking habits. Red-footed boobies (**6**), tropical equivalents of gannets, commonly breed on the Galápagos Islands, though better-known inhabitants are the seaweed-eating marine iguanas (**1**), the only lizards among 3,000 species to be truly at home in the sea.

4

5

ground finch, have powerful crushing mandibles to crack open tough seeds. The woodpecker finch has adopted a tool-using lifestyle, probing crevices in trees and rocks with a cactus spine held in its beak in order to extract the soft-bodied invertebrates that shelter within.

An even more spectacular example of adaptive radiation exists in the Hawaiian archipelago, which lies 3,200 kilometres from North America, the nearest continent. The western islands are considerably older than the Galápagos, being about sixteen million years old, although the larger islands at the other end of the 2,650-kilometre chain, including Hawaii and Maui, rose from the sea only about 700,000 years ago. The honeycreeper bird family is confined to the Hawaiian islands. All members are fundamentally similar, suggesting that they are derived from the same ancestral stock, but, like Darwin's finches, they have evolved a multitude of different feeding adaptations which, allow them to exploit the huge range of vegetation that occurs on these islands.

6

The ancestral form of the honeycreeper is thought to have had the broad beak and muscular tongue typical of seed-feeding birds, but, in the absence of competition from other passerines, there are now some twenty-three species which have expanded to fill a wide variety of feeding niches. Some, such as the akohekohe, have slender beaks as an adaptation to preying on insects, and others have developed tubular tongues with which to sip nectar from flowers. Of these, the apapane and amakihi feed from shallow flowers and so have relatively short beaks, but the iiwi has an enormously long, downcurved bill to enable it to reach into deep, cylindrical blooms. Another species has crossed over mandibles like those of the seed-

all descended from the same ancestor, but have evolved into separate species following the isolation of the valleys during erosion of the island's volcanic base.

As a general rule, the total number of species present on an island decreases as distance from the mainland increases; but the level of endemism, as is to be expected, rises with the degree of isolation. Both factors are also dependent upon the age of the island or archipelago, its origins, size and the diversity of its habitats. Madagascar, for example, is only about 400 kilometres from the African shore, and is the fourth largest island in the world, with habitats ranging from desert to subtropical rainforest. It has also been an island for over 150 million years. It is perhaps not

Endemic birds of the Galápagos archipelago include the small ground finch (**2**), probably similar to the ancestors of Darwin's finches that were blown over to the islands from mainland South America, and the highly predatory Galápagos hawk (**5,6**).

feeding crossbills of the northern forests, and the Maui parrotbill has a tough beak for prising the bark away from trees to expose grubs and ants beneath.

Madagascar supports twelve species of vanga-shrike, which have also evolved from a single ancestor. They display much more physical variation than the honeycreepers or Darwin's finches, ranging from the hornbill-like helmet bird to the sickle-billed and hook-billed vangas. The brightly coloured kingfisher-like todies are endemic to the West Indies, and the same species of parakeet is represented by a yellow-headed race on Curacao, orange-capped birds on Bonaire and individuals with olive-green heads on Aruba.

What is immediately obvious from these examples is that, as a result of adaptive radiation, remote islands are often home to a large number of species that are not found anywhere else in the world; they are endemic to that island. But adaptive radiation and endemism are by no means confined to the bird kingdom. Among the plants of Hawaii, for example, no less than ninety per cent of the 2,220 taxa are endemic, and it has been estimated that more than 1,700 of these are descended from only 275 different ancestors. Also in the Hawaiian chain, crickets, fruit-flies and carabid beetles have undergone spectacular radiations. The 3,700 or so types of insect present on the islands are similarly thought to have evolved from only 250 original species. Five neighbouring valleys on the Hawaiian island of Oahu contain five different species of snail. These are

surprising, therefore, to find that Madagascar houses a total of some 12,000 species of vascular plant, about eighty per cent of which are endemic, making it one of the richest botanical areas in the world. In addition, about ninety per cent of Madagascar's indigenous reptiles are endemic and not one of its sixty-six mammals is found elsewhere in the world.

An even larger island, Borneo, is similarly distanced from Asia, but, because it was linked to the mainland by falling sea levels during the Ice Ages, it has a much lower proportion of endemic species than Madagascar. On the other hand, it supports over 600 different types of bird and no less than twenty-one species of primate, ranging from the tiny tarsiers and tree shrews to the orang-utan. Nearby Sulawesi, however, which remained isolated at this time, has a much more impoverished biota, but is richer in endemic species. On St Helena, in the South Atlantic, one of the most remote inhabited islands in the world, ninety-seven per cent of the plants are found nowhere else, whilst on the isolated volcanic islands of the Samoan archipelago in the South Pacific,

Today, the only surviving prosimians are the potto and bushbaby in Africa, the loris in Asia and some twenty-nine species of lemur on the island of Madagascar. The absence of large carnivores on this island made it an ideal haven for lemurs, and, what is more, their main competitors never arrived on the island – probably because Madagascar was so far away by the time monkeys became widespread.

The more typical lemurs include ring-tailed and ruffed lemurs and such monkey-like creatures as the sifaka and indri. But the most bizarre is undoubtedly the aye-aye, commonly described as an apparently hastily

5

6

8

The Madagascar periwinkle (**10**), also native to India, has gained fame as a source of alkaloids with valuable medical properties. The tomato frog (**5**) is one of Madagascar's more striking endemic amphibians, but the most characteristic denizens of the island are the lemurs, ranging from the so-called typical lemurs, such as the black (**9**) and ring-tailed (**6**) species, to the noisy, tailless indris and the extraordinary, nocturnal aye-aye (**8**). The stocky, nocturnal Tasmanian devil (**1**) is the second largest marsupial carnivore in the world. The white (fairy) tern (**4**) lays a single egg perched precariously on a branch, and the chicks barely move for fear of losing their balance.

7

9

thrown together assortment of parts left over from other animals. Its teeth are distinctly rodent-like in that they never stop growing; it has a bushy tail reminiscent of that of a fox; ears like a bat and, instead of the finger and toenails of other lemurs, it has claws. But the most peculiar feature of the aye-aye is its middle finger, which is almost skeletal in its lack of flesh and fur. During their solitary travels through the forest, aye-ayes have been seen and heard tapping on the wood of ancient trees with this skeletal finger, presumably to see if they are hollow and likely therefore to harbour invertebrates. Then, using its sharp incisors, it will gnaw a neat hole in the wood, through which it can hook the animals out of their refuge.

Madagascar is also the stronghold of another ancient mammal group: the tenrecs. These are the descendants of primitive insectivores that have remained

10

almost unchanged over 100 million years and are found nowhere else in the world today. There are twenty-one species of tenrecs on Madagascar, some of which look very much like European hedgehogs, complete with an armoury of spines. Some resemble the mole in both looks and deeds, whereas others are more shrew-like. The solenodon is another primitive insectivore, confined to a few islands in the West Indies. Like the tenrec, it has changed little since it first appeared on earth, but there are only two species of it left in the world today, one confined to the island of Cuba, and one to the Haitian end of Hispaniola. Between forty-five and fifty-eight centimetres long, they bear a superficial resemblance to large shrews, although, unlike these harmless insectivores solenodons are highly venomous. Poison is produced in glands under the jaw so that when the solenodon bites its prey – small vertebrates and large invertebrates – the venom flows along grooves in its lower incisors into the body of the victim.

Across the world's islands there are many examples of animals and plants that are miniatures of their kind. It has long been accepted that an island is only able to harbour a certain population size of any one species within its strict boundaries. But if the individuals of this species are smaller, then the number that the island can support is theoretically higher, the population has greater genetic variation and is thus more stable. On Tenerife, the largest of the Canary islands, one type of

An attractive shieldbug (**1**) sits on a leaf on the volcanic island of Krakatoa. Such winged insects are among the first forms of animal life to arrive on sterile, newly-formed islands, although they do not always become permanently established. The Komodo dragon (**2**), which, at around three metres, is the world's largest lizard, is thought to have attained such a size owing to the absence of competition from large mammalian carnivores. Like all monitor lizards, the Komodo dragon is able to dislocate its lower jaw in the manner of some snakes to allow it to swallow large prey such as monkeys and deer.

lizard is almost twice as long as the same species on Hierro, the smallest island of the chain. Similarly, the mouse lemur of Madagascar is the smallest primate in the world, averaging only ten centimetres in length, and even pygmy elephants were once common on some of the Mediterranean islands and in Indonesia.

At the opposite extreme, however, are the island dwellers that have become veritable giants, probably as a result of the lack of competition for food. There are giant grasshoppers in New Zealand, centipedes up to thirty centimetres long in the Galápagos archipelago, and earthworms three centimetres in diameter and three to four metres long in Australia: the bulkiest in the world, if not necessarily the longest. The Komodo dragon is the largest lizard alive today. Discovered on the Indonesian island of the same name in 1912, it is a type of monitor lizard, the males of which commonly exceed three metres in length and weigh up to 100 kilograms. It is a voracious carnivore, hunting by day and locating its prey mainly by sight; there are no other large carnivores on these islands, which may partly explain how it has been able to reach such a size.

The giant tortoises are among the best known island denizens in the world. There are fifteen known subspecies of giant tortoise in the Galápagos, four of which are probably extinct and another that has only one known living representative. For the most part

these species, all of which are rare today, exist on separate islands of the archipelago, although the largest island, Isabela, contains five different subspecies, each effectively isolated from the other by high volcanic ridges. Some races may grow to as much as one and a half metres in length, weigh up to 270 kilogrammes and are thought to live up to 200 years – longer than any other terrestrial animal.

The Galápagos tortoises almost certainly originated in South America, but another giant species evolved independently on the scattered islands of the Indian Ocean – Mauritius, Réunion and Rodrigues as well as the Comoros Islands and Madagascar – although today it is found only on the remote atoll of Aldabra, which lies about 600 kilometres off the coast of Africa. This species is thought to have originated in Africa, but developed its large size when it was cut off from its mainland relatives on Madagascar. From Madagascar, colonisation of the other islands would not have proved difficult, owing to the fact that giant tortoises float and can therefore survive long periods at sea. The direction of the Indian Ocean currents today would carry a gravid female to Aldabra from Madagascar in less than two weeks.

Among the many island plants are the giant silverswords of the Hawaiian islands, similar to the giant lobelias and groundsels of the East African high

The huge and venerable Galápagos tortoises (**3**) show distinct variations across the archipelago in the shapes of their carapaces, allowing the identification of no less than fifteen different subspecies. Each is, or was before it became extinct, confined to a different island, although Isabela boasts no less than five subspecies, each separated from the others by high volcanic ridges that have ensured their evolution in isolation.

mountains. They take fifteen years to reach maturity and then produce a two-metre flowering stem. In the absence of more typical tree species, the Galápagos Islands possess forests of giant prickly-pear cacti and fifteen-metre shrubs, often known as tree-daisies. St Helena, in the South Atlantic Ocean, has no less than five very different species of tree-daisy, some of which are seven metres tall. An eight-metre-tall tree of the herbaceous bedstraw family can be found growing on the Polynesian island of Samoa. The Canary Islands have large numbers of giant buglosses, which are similar in appearance to the Hawaiian silverswords, whilst the largest flowers in the world, more than sixty centimetres in diameter and belonging to a genus of parasitic plants known as Rafflesia, are found on the island of Borneo.

The moas of New Zealand were enormous birds which, in the absence of mammalian herbivores, were able to take full advantage of the vegetation of these islands. The largest of these measured three and a half metres from head to toe, and may have fulfilled a similar role to the giraffe's in browsing among the tree

4

6

5

7

8

9

Of the seventeen species of penguin in the world, all but one live in subantarctic and Antarctic regions, including the diminutive rockhopper (**7**) and larger gentoo penguins (**9**), both of which breed on Steeple Jason.

King penguins (**4**) lay their eggs in spring or early summer, the male balancing the single egg on its feet beneath a warming flap of abdominal skin. The chicks (**5**,**6**) soon develop a thick, woolly brown coat and are fed throughout the winter, becoming independent the following summer, although they don't breed until they are six years old.

tops. Likewise the elephant bird of Madagascar was the bulkiest bird ever to have existed, at over three metres tall and weighing in at around 450 kilogrammes. Its eggs had a capacity of more than nine litres. But apart from such huge size, these birds had another distinguishing feature: like the emus and ostriches of today, they could not fly.

This loss of flight is observed again and again among the birds of remote islands, and, indeed, the insects. Even the parachutes and wings of plant seeds become reduced. One theory is that having reached a suitable island where predators and competitors are rare or absent altogether, flight becomes a disadvantage to an animal. In terms of energy consumption, flight is very expensive, especially the effort of becoming airborne. Even maintaining the wings and their associated muscles (which make up about one fifth of the weight of a typical bird) proves costly in terms of the extra nourishment required. With only a limited amount of resources available, it makes sense to dispense with such unnecessary apparatus, but in doing so these creatures seal their own fate, since their descendants can never again disperse to pastures new.

share of flightless birds, including the world's only species of flightless cormorant, which is confined to the rocky coasts of Isabela. The white-throated rail of Aldabra has only vestigial wings, but the same species retains the power of flight on Madagascar. Flightless rails are also present on Tristan da Cunha, Ascension Island and Gough Island in the Atlantic Ocean.

Island animals and plants are particularly vulnerable to extinction, partly because of the small size of their populations and partly because of their restricted range – some are confined to just one small valley on a remote oceanic island. Of the ninety-four species of birds that are known to have become extinct since the year 1600, only nine were continental in origin, despite the fact that only about twenty per cent of the world's birds live on islands. Unfortunately, such creatures are threatened from many sides, particularly by natural catastrophes and the ever increasing influence of man.

The very nature of many oceanic islands renders their wildlife susceptible to sudden extinction since a good proportion are volcanic in origin and the forces which squeezed them from the sea bed may still be

Cape gannets (**1**) form immense breeding colonies on selected islands off the South African coast; the harvest of the resulting guano is one of the richest in the world. Black-browed albatrosses (**6**) are also colonial breeders, often building their mounds of earth and grass on windy slopes since they need to be able to run downhill to become airborne. Swallow-tailed gulls (**3**) breed mainly on the Galápagos Islands.

1

New Zealand is a good place to observe flightless birds since, of the fifty or so species of land birds which are endemic to these islands, some fourteen are either poor fliers or completely flightless – including a species of rail known as the weka. There are three species of kiwi, possessing such long, hairlike feathers that their tiny, useless wings are virtually invisible. The largest parrot in the world, the kakapo, weighs over three and a half kilogrammes, so instead of flying, this bird prefers to clamber through the branches using its wings only for balance. The takahe, a flightless coot the size of a turkey, with a massive red bill and metallic blue plumage, was thought to be extinct for many years until a small colony was discovered in 1948 in a remote South Island valley.

Among Australia's 650 indigenous bird species are the flightless emus and cassowaries, the latter confined to tropical forests in the north and New Guinea. On New Caledonia, to the east of Australia, a splendid white, crane-like bird called the kagu has also found wings to be an unnecessary burden during the course of its evolution. The Galápagos archipelago also has its

2

Although it lies almost 1,500 kilometres from the coasts of both Africa and South America, immature birds from both continents, such as waxbills (**2**) and black-crowned night herons (**5**) respectively, frequently turn up on Ascension Island, presumably blown off course during their first migration flights.

Members of the auk family are the northern hemisphere equivalents of the penguins, although they still retain the power of flight. Guillemots, known as California murres in the North American part of their range, even hold their single, pear-shaped egg on their feet to keep it warm during incubation, as do king and emperor penguins. The European black guillemot (**7**) is the least sociable of all auks, nesting in cliff-face crevices, while the smaller pigeon guillemot (**4**) of the Pacific coast of North America prefers rocky overhangs near sea level.

operative. The devastating eruption of Krakatau in 1883 eliminated all life on the island, and the resultant tsunami killed more than 36,000 people living in coastal villages on nearby Java and Sumatra. Kilauea and Mauna Loa on Hawaii are active volcanoes , although Mauna Kea is thought to be extinct. Tristan de Cunha in the South Atlantic is also still growing. Its last violent phase of volcanic activity occurred in 1961, but Teide on Tenerife in the Canaries last erupted in 1939 and thus presents less of a threat to the wildlife of the island today.

The eruption of Mount Soufrière in 1979, followed by Hurricane Allen in 1980, reduced the population of the endangered St Vincent parrot in the Caribbean by a quarter, but even small-scale or localised eruptions may be devastating to particular species. The Niuafo'ou

megapode is Tonga's only endemic land bird; its numbers are today estimated to be between 200 and 400. Because of the peculiar habit this species has of laying its eggs in loose soil near warm volcanic ducts to incubate them, even a small rise in temperature of the molten rock beneath the island would be sufficient to eliminate the progeny of an entire year.

Other natural catastrophes include earthquakes, such as that which destroyed Kingston, Jamaica, in 1907. The effect of hurricanes on island biota is not to be underestimated either. Because oceanic islands are usually surrounded by large expanses of sea, hurricanes and cyclones are able to build up massive speeds as they travel across the level ocean, striking with horrific force when they do encounter land. Only recently several of the Caribbean islands were laid waste by a series of hurricanes, and the wildlife of nearby Tobago was decimated by Hurricane Flora in 1963. Hurricanes, as well as providing the raw materials for island colonisation, also wreak great destruction.

The vagaries of the world's geological and climatic forces, however, are part of the natural order of life on earth. What is more disturbing is the effect that man has had on islands and their wildlife over the years. The earliest human colonisers of the world's various islands hunted the indigenous animals for food; a natural case of predator and prey, except that most of the species man hunted had evolved in the absence of large carnivores and therefore had not developed the necessary escape responses which preserve many continental animals from extinction. The arrival of man on Madagascar in about the fifth century resulted in the extermination of all the island's eleven species of giant lemur, some as large as gorillas.

Many large flightless birds were also wiped off the face of the earth by early seafaring folk and island settlers, who were delighted by the easy source of meat that such huge creatures provided. When the Polynesians arrived in the eighth century, New Zealand had no native mammals except for two species of bat and several seals. Instead there were some nineteen species of moa, which were quickly hunted into oblivion by these ancestors of the Maoris. On Madagascar a similar situation has resulted in the extinction of the elephant bird within the last 1,000 years. Likewise, the dodo of Mauritius and two similar birds called solitaires from nearby Réunion and Rodrigues vanished shortly after their discovery by seafaring races. In each case they displayed a remarkable lack of fear of man and, in any event, were unable to take to the air to escape. The last dodo was killed in 1681, less than 200 years after it was discovered, and the solitaires had succumbed to similar hunting pressures by the end of the eighteenth century.

Male frigatebirds develop bright red, inflatable throat pouches (**1**) during the breeding season, probably used both in attracting a mate and warning off rivals. Young chicks of unrelated seabirds look very similar in their white, downy coats, as can be seen from this white-tailed tropicbird (**3**) and black-browed albatross (**4**).

Giant tortoises were once very common, both in the Galápagos archipelago and the islands of the Indian Ocean. But early sailors soon discovered that these huge reptiles could remain alive on board ship for long periods of time, providing the sailors with a valuable supply of fresh meat. It has been estimated that some ten million Galápagos tortoises were killed in this way over the space of only a few hundred years, and by the end of the nineteenth century the Indian Ocean species had become extinct everywhere except for the remote atoll of Aldabra, which was far from the main shipping routes of that time.

Another very destructive custom practised by early seafaring races was that of leaving domestic animals on uninhabited islands to multiply, thus assuring the sailors of a decent meal when they next passed by. But the indigenous plants of such remote islands had, as often as not, evolved in the absence of large herbivores,

losing, in the process, such unnecessary traits as spiny foliage and toxins in their leaves, which normally serve to deter grazers. In many cases the natural vegetation of these islands was decimated in this way. For example the introduction of goats to St Helena in 1513 resulted in the extinction of forty per cent of the endemic plant species in the space of a few decades. Perhaps a more vivid example is that of Round Island near Mauritius, which is only 151 hectares in area but has the misfortune to lie on a major trade route. Goats and rabbits were introduced about 100 years ago, and all that remains of the once crowning hardwood forest today is a single tree.

Deliberate introductions of destructive domestic animals onto uninhabited islands was only the tip of the iceberg. Goats, cattle, sheep and poultry all accompanied man in his quest to exploit new lands, but several undesirable species, especially rats, were

Most species of *Pelargonium* (**2**) are native to South Africa, but their popularity as house and garden plants has led to their becoming naturalised on many islands inhabited by man, often at the expense of the indigenous flora.

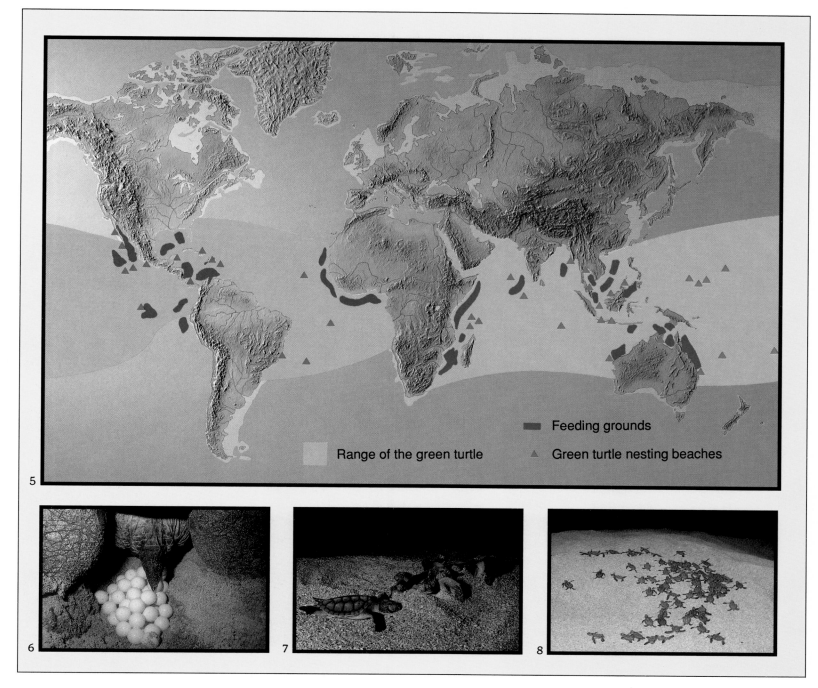

Range of the green turtle

Feeding grounds

Green turtle nesting beaches

5

6

7

8

unintentionally let loose in the process. Exotic plants and crops have since proliferated at the expense of the uncompetitive native flora. On the Hawaiian island of Oahu, eighty-five per cent of the native vegetation has been replaced by exotic plants, and although there is now strict control on the importation of plant material of foreign origin, it is too late to save the many endemic species that have already succumbed to the intruders.

The situation is so critical in New Zealand that it is often difficult to find a single native species today, and less than twenty per cent of the original forest cover remains. Since the arrival of the Europeans, no less than 133 non-native bird species have been introduced – domestic fowl for food, and cage birds to remind the early settlers of their home on the other side of the world – as well as sheep, cats, weasels, deer, trout and rats. Over a quarter of the introduced foreign birds are now established in the wild at the expense of seven New Zealand native species which have disappeared forever. In Australia, the primitive monotremes and marsupials are fighting a losing battle against the many placental mammals that have been introduced by human settlers, not least the dingo, which preys on the smaller herbivorous marsupials, and rabbits and sheep, which compete for foodstuffs with several of the larger kangaroo species.

In the Galápagos Islands, the overgrazing of the native vegetation by domestic livestock has severely reduced the degree of cover available to the indigenous land iguanas, which are consequently suffering from increased attacks by birds of prey. The related marine iguana, the only truly ocean-going lizard in the world and found solely on these islands, is declining following the attacks of feral predators, such as rats and cats, on its eggs and young.

Mongooses have been widely introduced throughout the Caribbean archipelago as a means of controlling other non-native species, such as rats. The success of this operation as an exercise in pest control was so great that the same procedure was applied to several of the Hawaiian islands. Unfortunately, rats are not the only victims of the mongoose; the Cuban solenodon is now an endangered species, the Jamaican and Puerto Rican boas are succumbing to these voracious carnivores, and Newell's manx shearwaters are now confined to Kauai – the only island in the Hawaiian chain from which this introduced predator is absent.

The Hawaiian islands were discovered by the Polynesians in the fourth century. Between that time and the arrival of white settlers about 200 years ago, it has been estimated that about half the avifauna became extinct, and of the sixty-nine endemic birds remaining at the time of the second human invasion, twenty-five have since vanished and most of the others are rare or endangered. The main reason for the recent wave of extinctions is thought to be largely due to the

Green turtles (**6,7,8**) live mainly in shallow, tropical waters, returning to the same beaches to lay their eggs, though they do not necessarily breed every year. The female green turtle has an arduous journey up the shore to lay her eggs, excavating the hole with her hind flippers, and will return at approximately two-week intervals between two and seven times. The young hatch simultaneously after forty to seventy-two days according to latitude, breaking out of their shells with the aid of a horny egg-tooth on the upper jaw, making their way immediately down to the sea, where they remain close to the shore, feeding on marine invertebrates for several months before heading for the open sea.

accidental introduction of two diseases associated with domestic fowl – avian malaria and bird pox – although feral cats and mongooses have also played their part.

Direct habitat destruction by means of timber extraction and clearance of native forests for agriculture have also had a devastating impact on island biota worldwide. Since Mauritius was colonised by man in the mid-1600s, eleven of the twenty-one endemic birds have disappeared, along with two thirds of the endemic reptiles. Today, the island is considered to possess one of the most endangered floras in the world, with many of its native species already having vanished forever. The main agent of destruction has been the exploitation of the native forests of Mauritius,

1

2

both locally for fuelwood and house construction, as well as commercially for timber and shipbuilding. Vast areas have also been cleared for agriculture. The black paradise flycatcher of the Seychelles once occupied the whole archipelago, but is now confined to those parts of La Digue where the forests have not been felled; whilst the Seychelles owl, once thought extinct, now occurs only in the mountain forests of Mahé Island. Similarly, the monkey-eating eagle of the Philippines has almost disappeared with the loss of its primary rainforest habitat, and the orang-utan of Borneo is declining following widespread logging in its habitat.

Madagascar is considered to support one of the richest assemblages of plants in the world today, but how much greater must it have been before large tracts of forest were cleared and eighty per cent of the native vegetation was destroyed? The inhabitants of the island today include some eleven million people and about ten million zebu cattle. The population is rising at a rate of three per cent annually and, to cope with the increased demand for food, about a quarter of the island is burnt to provide new pasturage every year. The Indonesian island of Java is one of the world's most overcrowded islands, its human population having increased from twenty million to more than ninety million inhabitants over the last century; a fact which was no doubt indirectly responsible for the extinction of the Javan tiger and the near elimination of the Javan rhinoceros during the eighteenth century.

Islanders struggling to rear domestic livestock will also kill any creature which they consider to be threatening their animals. The predatory kaka and kea parrots of New Zealand have suffered greatly at the hands of farmers, and both are staring the prospect of extinction in the face. Besides being persecuted by man because it was believed to carry off poultry, its nests are also frequently raided by introduced rats, and its habitat destroyed. It is hardly surprising that the Mauritius kestrel once bore the title 'rarest bird in the world', although the situation has improved a little today: their population has increased to number about twenty birds.

3

4

Cliff-nesting seabirds that favour offshore stacks and islands include tufted puffins (2) in North America, king shags (3) on the Falklands and kittiwakes (6) throughout the northern hemisphere. Waved albatrosses (7) prefer more level ground.

The Falkland Islands are home to an endemic species of steamer duck (9) similar to that which inhabits the South American mainland, except that it has lost the power of flight. Black-crowned night herons (5) also drop in occasionally.

5

Because the Komodo dragon, the largest monitor lizard, often takes domestic livestock, it has been heavily persecuted in the past. However, apart from man's detrimental influence, it has been suggested that this lizard is not adapting well to recent climatic and environmental changes in the area. It is thought that, as with crocodiles, temperature strongly influences the sex of the developing embryos. The sex ratio of the population today is strongly biased towards males: of a total of some 2,000 Komodo dragons only about 400 are females. It may be that the Komodo dragon will be one of the few creatures on this planet to become extinct from natural causes rather than at the hands of man.

A more recent threat to island ecosystems hinges on the increasing amount of leisure time available in the mechanised Western world combined with the lure of the inherent beauty of many tropical islands – tourism. Of the 1,500 coral islands of the Maldives in the Indian Ocean, a mere 200 are occupied by fishermen, but previously uninhabited islands are sprouting hotels and leisure resorts as if overnight, which will surely have a devastating impact on the area's wildlife. In this part of the world, tourism is a fairly recent innovation, but many of the world's islands have been suffering seasonal influxes of suntan oil and beach towels for decades, especially those in the Mediterranean, Caribbean and Bahamas.

The Galápagos Islands attract visitors of another type: those who are keen to see the remarkable assemblages of unique animals and plants of the

6

7

8

Ascension Island has lost the immense colonies of breeding seabirds it once housed following the accidental introduction of rats, and the subsequent deliberate introduction of cats to control them. Only wideawake terns (**8**) still breed here in large numbers. An equilibrium has been established as the terns stay out at sea for nine months of the year, causing the predator populations to crash because of lack of food during this time. Even so, only those chicks located in the centre of any colony are likely to survive to maturity. Other seabirds which still breed on Ascension Island are white boobies (**1**) and red-billed tropicbirds (**4**), although in greatly reduced numbers.

archipelago for themselves. And yet this 'green' tourism is adversely affecting the very features which these visitors have come to admire. The friable volcanic slopes of the islands are trampled and eroded, and alien seeds, attached to clothes and shoes, spring to life to threaten the indigenous vegetation. In the early 1980s, a limit of 12,000 visitors per year was agreed upon by the Government of Ecuador and the Galápagos research stations, but this was soon exceeded, and in 1986 over 30,000 people visited the islands. This is considered to be far in excess of what the archipelago can support, and moves are afoot to reduce the annual maximum to 25,000.

The lack of foresight, which has characterised man's attitude to the islands of our world in the past, has caused many forms of life to vanish for ever from the face of the earth. But the very fact that we are aware of such losses may mean that we have learnt a valuable lesson in dealing with such fragile and unique island ecosystems, a lesson which we can perhaps apply to other aspects of our existence. For above all, the planet earth itself is an island, vulnerable and utterly isolated in the vast eternity of the universe, and we must learn to treat it with respect.

9

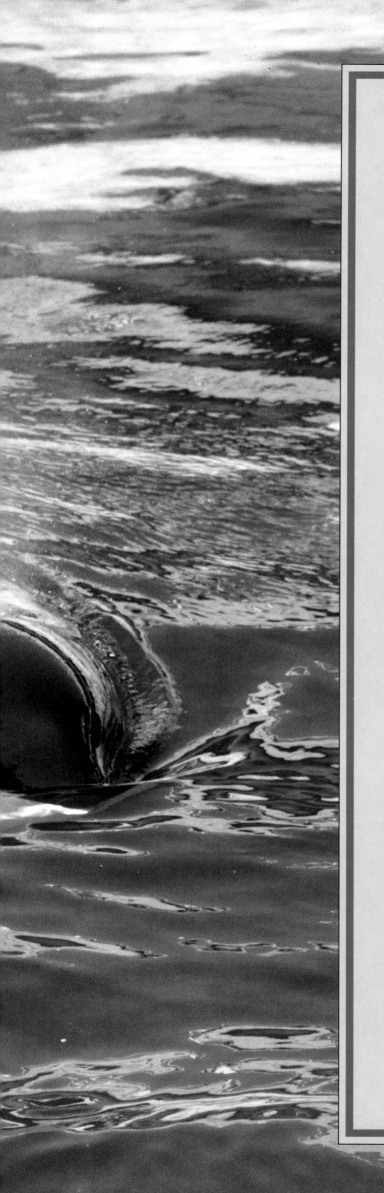

· CHAPTER 7 ·

OCEANS

Almost three-quarters of our planet, despite being called Earth, is in fact concealed beneath a vast expanse of ocean. From the sunlit waters of the surface to the cold darkness of the abyssal plain, the salty oceans are teeming with life which can exist nowhere else. The ocean deeps remain largely unexplored by man, so there are undoubtedly creatures existing in this realm of permanent darkness which have yet to be discovered. For centuries man has been exploiting the resources of the seas and emptying the debris of his terrestrial existence into the ocean without understanding the laws by which it operates. In recent years, however, fish stocks have declined, mysterious diseases have struck down seabirds and marine mammals alike, and the great whales have been hunted to the brink of extinction. It is only now that the vulnerability of the seas and their unique forms of life is becoming evident.

Left: Orcas, justifiably known as killer whales. Top: peaceable, plankton-feeding humpback whales and (above) the coelacanth, discovered in 1938.

Of the total surface area of the Earth, the oceans account for more than seventy per cent, separated by the continents into the Indian, Atlantic, Pacific, Arctic and Antarctic (Southern) oceans. Many subsidiary seas also exist, some of them almost completely enclosed by land, such as the Mediterranean, Baltic, Red and Caribbean seas, but all are connected with the main oceans, forming one continuous ecosystem – the largest in the world.

Though the average depth of the world's oceans is about 4,000 metres, their topography is incredibly varied. Great trenches and mountain formations occur along the peripheries of the main continental plates, deeper and higher than their terrestrial equivalents. The Marianas Trench in the North Pacific, at 11,022 metres below sea level, represents the deepest known point on the sea bed, while the volcano of Mauna Kea in the Hawaiian Islands exceeds 10,000 metres from the ocean floor to the summit, which lies above the surface of the water.

The continental shelf is so called because not long ago it was part of dry land, but it has gradually become flooded as water frozen during the ice ages melted and caused the sea level to rise. It is defined by its gentle gradient (often barely noticeable on land), from just over one kilometre off the western coast of the Americas and around Africa to about 900 kilometres south of Newfoundland. Beyond the continental shelf lies the continental slope – distinguished by a sudden increase in gradient – and the boundary between the two is in effect the true edge of the land mass. Across the world the maximum depth of the sea at this changeover point ranges from a mere ninety metres in the Mississippi Delta to over 700 metres off Florida. The continental slope descends to the ocean floor, otherwise known as the abyss.

The main factors influencing life in the oceans are light, salinity, temperature and pressure – all of which are ultimately dependent upon depth. About ninety-nine per cent of the ocean is completely dark. Much sunlight is dissipated by reflection at the surface and the remainder is absorbed within a few hundred metres. The depth to which there is sufficient light available for plants to photosynthesise – the euphotic zone – varies with the clarity of the water; in tropical waters it may exceed 200 metres but in the northern seas it rarely extends beyond forty to fifty metres. Immediately below this zone light still filters through the water, but it is the wrong wavelength for photosynthesis, and deeper still lies the ever-dark aphotic zone.

Pure water accounts for about ninety-six-and-a-half percent of the total volume of sea water. The remainder is composed largely of dissolved salts, more than threequarters of which is sodium chloride. Of the earth's 103 naturally occurring elements, eighty-four

The vast oceans of the world are teeming with animal life, ranging from the primitive medusas (7) and sea anemones (4) to the majestic killer (3) and southern right whales (1).

2

3

4

The killer whale's distinctive black and white coloration possibly helps parents and their young to recognise each other. Measuring two metres at birth, the calf (3) remains with its mother for more than a year.

5

Moray eels (**6**) are voracious underwater predators that are largely confined to tropical waters. They occupy holes in reefs and underwater cliffs, launching lightning attacks on unsuspecting passers-by. The wandering albatross (**5**), with its three-and-a-half-metre wingspan, the largest of any species, is able to glide for hours over the open ocean in search of food, expending only the minimum of energy.

6

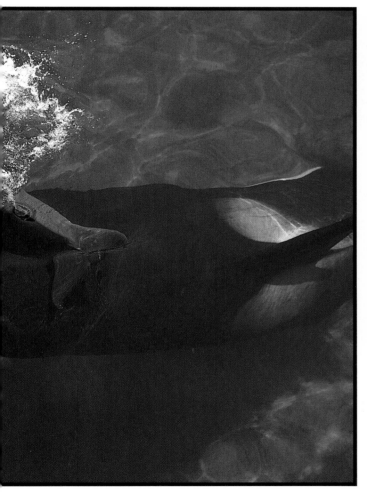

7

are present in the sea, although the majority are present only in very low concentrations. It is these lesser nutrients – especially nitrogen and phosphorus, essential for life but present in minute quantities – that determine the maximum size of marine populations. Generally, the proportions of the major chemical constituents of sea water are more or less constant, normally falling between thirty-two to thirty-eight per cent.

Deviations from this norm are dependent upon the evaporation rate and the addition of fresh water from rivers or melting icecaps. The Baltic Sea, for example, is almost completely surrounded by land and lies only just south of the Arctic Circle. During the spring, when evaporation is low, the numerous rivers arising in Scandinavia and western Russia endow this sea with the lowest recorded salinity in the world; a mere five per cent. The Red Sea, on the other hand, by virtue of its proximity to the equator and constricted connections with the Mediterranean and Arabian Seas, may suffer from salinities approaching forty-three per cent during the summer.

The temperature of the oceans depends on exposure to the sun and thus gradually decreases from the surface to the depths and also from the tropics to the poles, since at the equator the sun's rays strike the water almost perpendicularly and thus penetrate further. On the ocean floor near Antarctica the water temperature is constant at around -2° C, the lowest in the world, increasing to a maximum of some 35° C at the surface of some enclosed tropical seas such as the Red Sea and the Persian Gulf. But once away from the surface waters, regardless of latitude or season, there is little variation; between 1,800 metres and 3,000

The chambered nautilus (**2**) is a primitive cephalopod which scours the ocean floor or reef for small fish and invertebrates, trapping its prey with short, mobile tentacles. Mature individuals may possess more than thirty chambers in their shells, although the animal occupies only that most recently formed. The remainder are filled with gas, the quantity of which can be adjusted, allowing the animal to vary the height above the sea bed at which it operates.

metres the temperature is a constant 4° C in oceans all over the globe. In contrast to temperature, light and oxygen – all of which decrease with depth – pressure increases by one atmosphere for each ten metres descended, so that at 4,000 metres – the average depth of the oceans – the pressure is an incredible 400 atmospheres.

Oxygen content and salinity are fairly constant throughout the world's oceans due to the actions of currents, waves and tides, all of which contribute to the mixing effect. Tides, created by the gravitational forces emanating from the moon and the sun, affect mainly coastal areas, and although the circular churning actions of wind-generated waves stir the surface waters rhythmically, their effect is rarely felt a hundred metres below. The most important mixing motions are caused by global ocean currents, the pattern of which is closely correlated with the circulation of the earth's atmosphere; they move in a clockwise direction in the northern hemisphere and anticlockwise south of the equator. Such currents are the result of constant strong winds over the open ocean and forces created by the rotation of the earth. They are so efficient at preventing stagnation of the oceans that, even at the deepest levels, oxygen depletion is rare.

Warm water and that with a lower salinity is 'lighter' or less dense than cold or very salty water, and thus will lie nearer to the surface. Warm and cold 'streams' often move in opposite directions, such as at the Straits of Gibraltar, which link the North Atlantic to the Mediterranean. Where the prevailing winds drive the surface water away from the shore, cold water from the depths of the ocean rises to take its place. Such conditions are typified by the Humboldt 'upwelling', which occurs along the west coast of South America, and the currents rising around the edges of Antarctica. These currents bring up nutrients from the sea floor which, combined with the increased oxygen-carrying capacity of the cold waters, create areas of very high productivity for marine life.

Our planet is thought to have formed from a ball of incandescent gases some 4,500 million years ago. As the gases cooled and became solid, hot water vapour condensed on the surface: the first oceans. At this time the seas were very dilute, since they were mostly the product of a global distillation process, but waters expelled from volcanic vents and geysers gradually added chlorides, bromides, nitrogen and other elements.

In this warm, primeval broth, possibly stimulated by lightning, it is postulated that the first life forms were engendered – single-celled organisms very similar to modern day bacteria and blue-green algae – and from this simple beginning evolved all the myriad life forms of the world today. Some groups of animals, such as echinoderms, corals, jellyfish and seaweeds, never graduated from the marine environment, but others, in the course of developing more complicated body structures and biochemical processes, moved into freshwater lakes and rivers and were eventually able to colonise land. It is with good reason that the ocean is referred to as 'the cradle of life'.

Plants were the first organisms to make the move from water to land, around 550 million years ago, modifying the barren terrestrial environment in such a way that it became suitable for higher forms of life. The first amphibians evolved some 350 million years ago and are still dependent upon the aquatic environment for part of their life cycle. Dinosaurs, the largest reptiles that ever existed (whose eggs, like those of all reptiles today, were independent of an external water source), reigned supreme for some thirty million years. The earliest types of bird evolved forty million years ago, by which time the first primitive land mammals were in existence.

Yet at the same time as these creatures were adapting themselves to cope with life on land, some

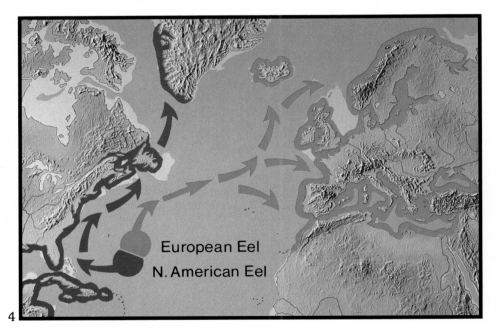
1
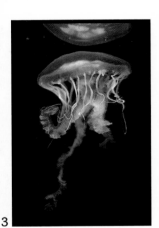
2

3

4

5

In the ocean depths the angler fish *Caulophryne jordani* (6) tempts its victims with a luminescent lure derived from the modified first ray of the dorsal fin, while *Astronersthes* (8) has huge jaws which dislocate, allowing it to swallow any morsel of food, even quite large pieces, which it might encounter. In the cool, oxygen-rich currents surrounding Antarctica, marine invertebrates such as *Antarcturus* (1) attain a phenomenal size.

European Eel
N. American Eel

Both European and North American eels spawn in the Sargasso Sea, and their leaflike larvae are carried by the Gulf Stream to estuaries on both sides of the Atlantic (4), where they change into needle-shaped elvers (5) before heading upriver to mature.

The coelacanth was known only from fossils until 1938, when the first specimen was hauled up by fishermen off the Comoros Islands in the Indian Ocean. It is thought to be closely related to the lobe-finned ancestors which crawled from the ancient seas to become the amphibians of today. The coelacanth is almost always encountered in the 180-200-metre zone, where the water is very cold, light is scarce and disturbance from surface waves is minimal: an environment which has almost certainly remained the same for millennia. The six species of nautilus, descendants of an ancient group of cephalopods – squid, octopus and related creatures – are found in depths ranging from 50-650 metres. They possess

The discovery of the coelacanth (7), a primitive lobe-finned fish thought to have been extinct for ninety million years, off South Africa in 1938 and the capture of many more in subsequent years has provided a unique opportunity to study what may be the missing link between fish and amphibians.

6

7

8

10

9

were already returning to the sea. Around 200 million years ago, large reptiles, such as ichthyosaurs and plesiosaurs, were finding an aquatic existence more to their liking and, although these creatures eventually became extinct, other reptiles are still found in the oceans today, namely turtles and sea snakes. The first mammals are thought to have returned to the sea some fifty million years ago; these were the ancestors of those present-day marine leviathans, the whales. Several million years later seals began to invade the oceans from the coastal regions, but they are not as fully acclimated to the marine way of life as the whales, since they still have hind limbs and return to the land to breed.

The marine environment is far more constant than its terrestrial counterpart, its inhabitants being sheltered almost entirely from changes in climate which affect the earth. For this reason, marine organisms have little reason to reform their lifestyle and many have remained almost unchanged since they first evolved. Today's sharks, for example, share many features with their fossil ancestors, which first appeared about 350 million years ago, and other animals, such as the coelacanth and nautilus, are little more than living fossils.

The pelagic realm is home to such bizarre cephalopods as the argonaut (2), with its wafer-thin shell, and the sea butterfly (10), its muscular foot extended into two mobile wings. In a region which offers little concealment from predators, transparency is a common phenomenon, as in the comb jelly (9). The jellyfish *Dactylometra quinquecirra* (3) possesses a bell-shaped canopy which can be flexed to provide lift, and trailing tentacles with which to trap planktonic organisms.

chambered shells similar to those of. the fossil ammonites. As the animal grows, new chambers are added, the vacated sections being filled with a mixture of gas and liquid. By regulating the amount of fluid present, the nautilus can adjust its buoyancy, and thus the level at which it lives.

Compared to the extent of the interface between land and sky which is available for terrestrial animal and plant life, the ocean presents a much more three-dimensional habitat'. Furthermore, plant life is not fixed in one place, but is mobile. Two main environments are available to marine organisms: the pelagic realm and the benthic realm. Pelagic organisms are those which are occupy the huge volume of the oceans, but benthic organisms are usually fixed to the sea floor, although for convenience many mobile bottom-dwelling creatures are also included in this category.

The inhabitants of the pelagic realm can be further divided into planktonic and necktonic organisms, according to whether or not they are able to move independently of ocean currents. Plankton are normally unable to choose in which direction to travel because they are so small that their frantic swimming motions have no effect against the powerful ocean currents. Nektonic creatures, on the other hand, are all vertebrates or cephalopods which have the necessary musculature to be able to navigate at will, selecting their own destinations.

There are two types of plankton: phytoplankton, which, like land plants, are able to harness the sun's energy to grow and reproduce by means of photosynthesis, and zooplankton, which cannot feed in this way, and instead consume the primary producers. Most phytoplankton are single-celled organisms such as diatoms, which abound in cold waters, or

Sea slugs resemble terrestrial slugs in that both are gastropods that have lost their shells during the course of evolution. Sea slugs, however, are also known as nudibranchs, meaning 'naked gill', since their feathery respiratory apparatus is carried externally on their backs. Many feed on sea anemones. The nematocysts with which the anemone traps its food somehow fail to discharge when eaten by the slug, and later pass intact through the wall of the intestine and migrate to the dorsal region, where they serve to protect the slug from predators. There is a phenomenal diversity of shape and colour within this group of marine creatures: (**1**) *Phidiana pugnax*, (**2**) *Triopha carpenteri*, (**3**) *Chromodoris macfarlandi*, (**4**) *Dirona albolineata*, (**5**) *Hopkinsia rosacea* and (**6**) the Spanish shawl *Flabellinopsis iodinea*.

dinoflagellates, which are more common in tropical seas.

Naturally all phytoplankton are confined to the euphotic surface layers of the sea by their need to trap sunlight. The numbers of phytoplankton in the seas vary with the availability of nutrients and oxygen, but there may be as many as 200,000 individuals in one cubic metre of sea water. Seas overlying upwelling areas and the continental shelf (because of the influx of nutrients from rivers) support rich phytoplankton communities, but warm waters are usually rather deficient in this respect. The Mediterranean, for example, contains three to ten times less phytoplankton than the Atlantic Ocean, which has repercussions for other marine life there, since all trophic levels are ultimately dependent on these oceanic 'pastures'.

Zooplankton can be further divided into the holoplankton, which are permanent inhabitants of this fauna, and the meroplankton, which consist of the eggs and larval forms of species which, as adults, are typically members of the free-swimming necktonic or sedentary benthic faunas. More than sixty percent of the holoplankton are crustaceans, especially copepods (common in Arctic waters), of which there are some 2,000 species, mostly grazers on the phytoplankton 'savannas'. Another major component of the holoplankton comprises the shrimp-like creatures called krill, of which there are some ninety species, found at

depths of up to 2,000 metres. Mostly they are about three-centimetres long, but in Antarctica, where stocks of krill are possibly the largest in the world, they may reach five or six centimetres.

Some zooplanktonic creatures, such as the jellyfish, are carnivorous, feeding on other zooplankton rather than the phytoplankton. Some jellyfish possess stinging cells in their tentacles called nematocysts which paralyse their prey before transferring it to a central mouth. Other predators include the tiny, transparent arrow-worms, which have sufficient muscular development to enable them to dart suddenly at creatures which stray within their range.

In order to maintain their position in the surface seas, planktonic organisms have evolved a number of features to assist them. The simplest of these involves the maintenance of the internal body fluids at more or less the same concentration as the salinity of the surrounding sea water, so that the organism will neither sink nor float. Some species of jellyfish, for example, are ninety-six to ninety-eight per cent water. They can also contract their 'umbrellas' somewhat to give them a slight lift. Other organisms compensate for a heavier body weight by containing air bladders; oil globules, as in the case of diatoms; or chambers full of water and carbon dioxide, as in some radiolarians. Alternatively, they attach themselves to larger, surface-dwelling organisms. Whales often carry profuse growths

The garibaldi (**10**) is a type of damselfish about thirty centimetres long found off the coast of California. Like all members of the family *Pomacentridae* it has a deep body and a forked tail and is aggressive and strongly territorial. Moray eels are large tropical and Mediterranean eels with naked, mottled skins and no lateral fins. They are often brightly coloured, as are these spotted moray eels (**11**), with a keen sense of smell and sharp, grasping teeth.

9

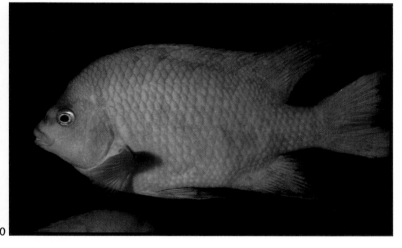

10

The scorpion fish (**8**) is a bottom-dwelling creature with a large number of spiny appendages, mostly derived from the fin rays, which are poisonous and also help to break up the outline of this fish for camouflage. The red gurnard (**9**), known as the sea robin in the United States, is a close relative of the scorpion fish. It has a strongly armoured head and the first few rays of its pectoral fins have been modified to act as feelers and to allow it to 'walk' across the sea bed.

11

of barnacles, algae, copepods, whale lice and remoras, whilst even turtles, large fish and ships' hulls are colonised by such 'hitchhikers'.

Swimming actions also help to keep planktonic creatures near the surface, and vary from the lashing of a single flagellum in the dinoflagellates to the rapid beating of legs and antennae (up to 600 times per second) seen in some copepods. The gastropod molluscs known as sea butterflies have also adapted to life in the surface waters: they have a foot which is expanded laterally into two lobes and is flapped slowly for swimming.

The Portuguese man-of-war and the velella, or sail jellyfish, are two larger members of the zooplankton which utilise gas-filled floats both to stay at the surface and also to travel to new feeding grounds by harnessing

1

2

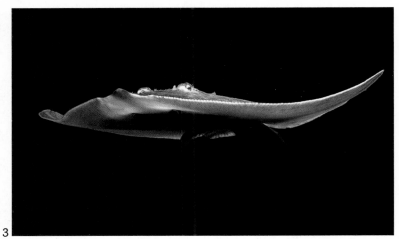

3

Octopuses (**2**) are cephalopod molluscs with an internal vestigial shell and eight muscular tentacles armed with suckers for grasping food, especially crabs and shellfish, or rocks for anchorage. They are common inhabitants of coral reefs, but are also found in many temperate seas and are renowned for their ability to change colour to match their background.

wind energy. The Portuguese man-of-war's float may be fifty-centimetres long and it must 'capsize' regularly in order to moisten its membrane. The largest true planktonic organism is a species of lion's mane jellyfish which is some two-and-a-half metres in diameter, has trailing tentacles over thirty-metres long and is found in the Arctic Ocean. Most planktonic creatures, though, are small, with a high surface-area-to-weight ratio that creates a degree of friction between the organism and the surrounding water molecules that ensures that it sinks only slowly. Phytoplankton with siliceous or calcareous 'skeletons' have developed spines, hairs or flattened appendages to increase this friction. Such apparently fragile creatures are able to exist in the oceans because of the cushioning effects of sea water, and the absence of the physical stresses with which land animals and plants have to contend.

Probably the greatest concentration of non-microscopic organisms in the surface waters of the oceanic realm occurs in the Sargasso Sea in the North Atlantic Ocean. Sargasso weed, also known as sea holly on account of its prickly leaves, has evolved into two separate, drifting species, complete with pea-sized air bladders to help them float. The weed provides shelter for large numbers of young fish and their predators, as well as for sea slugs and crabs – many of which have developed perfect camouflage to blend in with its cream and brown fronds.

The creatures which occupy the benthos vary considerably according to the type of substrate and the depth of the water. Shallow seas, where light penetrates to the floor, are often clothed in dense undersea meadows of eel-grass or seaweeds, which here add to the primary productivity of the food web. But where light at the sea bed is insufficient for photosynthesis and plants are absent, the bottom rung of the food chain is composed of organic detritus drifting down from the productive surface levels of the seas above. The only creatures present are those which can filter such particles from the water or extract them from the sediments, and these creatures' predators.

Soft, muddy substrata are the most extensive zones

4

of the sea floor, replacing the sandy and rocky bottoms which predominate along the shoreline. They are little disturbed by tides, waves and surface currents and thus support such soft-bodied animals as sponges, starfish and sea cucumbers on the surface; long-stalked sea pens, sea lilies, and polychaete worms which construct tubes so as to lift their delicate filter-feeding 'fans' clear of the sediments, and burrowing molluscs and segmented worms, which live within the mud itself. Some, such as sea anemones and sea fans, develop plant-like forms in which small animals shelter from the hunters of the sea floor in the absence of true vegetation. Many of the bottom-dwelling predators are flattened dorsoventrally, such as rays and guitarfish, while others are flattened from side to side, such as plaice and sole, the eye on the 'underside' gradually moving across the top of the head as the fish matures.

Like sharks, rays are cartilaginous fish, but differ from them in having gill slits beneath the head and flattened, winglike pectoral fins. Most of the 340 ray species are bottom-dwellers, such as the electric ray (**1**), but the small-eyed ray (**3**) and the huge manta ray (**5**) are both pelagic, the latter with a 'wingspan' of over six metres.

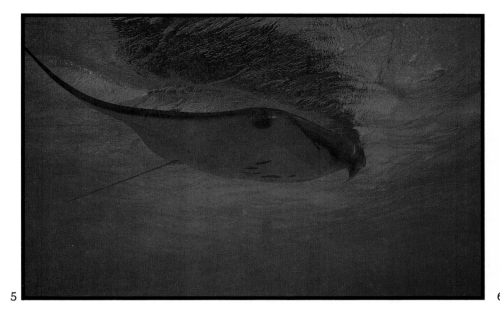

5

The great barracuda (**9**), found in tropical waters on both sides of the Atlantic and in the western Pacific, is among the fiercest of all piscine predators. It has powerful jaws and sharp, dagger-like teeth and may grow up to two-and-a-half metres long.

6

7

8

Man's fear of sharks is legendary, but in fact their bloodthirsty reputation is a vast exaggeration, perpetrated not least by the film industry. Biologists involved in shark research have found most species to be peaceable creatures, and do not hesitate to study them at close quarters (**4**). Some, however, do undoubtedly acquire a taste for human flesh from time to time. Blue sharks (**8**), for example, have been known to turn man-eater, and the sight of a dorsal fin will empty the water of bathers within seconds. The bizarre hammerhead shark (**6**) is also an aggressive species, but its small mouth limits the size of its victims. The largest of all sharks, and indeed of all fish, is the harmless whale shark (**7**); its huge, oily liver gives it sufficient buoyancy to remain at the surface, where the plankton on which it feeds is most abundant.

In order to remain undetected some of these undersea hunters are able to mimic the pattern and colour of the sea floor, while others bury themselves in the sediments to ambush their victims.

Of the nektonic predators, one type of animal stands out above all others as the perfect killing machine: the shark. Some 350 known species of shark inhabit the seas of the world today, together with a similar number of closely related skates and rays. Judging by the fossil evidence, sharks have changed little in the 300 or so million years since they first appeared; their sinuous, streamlined bodies, undershot jaws and formidable rows of razor-sharp teeth are much the same today as in those primeval seas. They are found throughout tropical and temperate seas from estuaries to the deep oceans, avoiding only the cold, polar regions.

Sharks have cartilaginous skeletons, which distinguishes them from the multitude of bony fish, or teleosts. In addition, instead of being covered with shiny, overlapping scales, a shark's skin bears tiny denticles, the texture of which is similar to coarse sandpaper. Unlike bony fish, sharks have no air-filled swim bladder for buoyancy, nor are they able to move their gill slits in order to pump water through and thus exchange dissolved oxygen for carbon dioxide. As a

9

consequence most sharks must swim almost continuously to avoid sinking and also to ensure that sufficient water passes through their open mouths in order to obtain oxygen – although certain slow-moving benthic species, such as the carpet shark, have overcome these problems to some extent. In addition, the pectoral fins of bony fish are highly mobile, used for changing direction and stopping, but those of sharks are fixed, so that they lack manoeuvrability, and in fact have no brakes.

Another difference between sharks and bony fish is that the latter release vast quantities of fertilised eggs into the sea when spawning, adding to the meroplankton, whereas about two-thirds of known shark species bear live young, and the remainder produce leathery, protective egg-capsules which are sometimes known as mermaids' purses. In species which give birth to live pups, the eggs hatch within the reproductive tract of the female, some of which are able to produce nutritive fluids to sustain the young, although in other species, such as the porbeagle, thresher and sand tiger, the first few to hatch turn cannibal and consume their siblings inside the 'womb'. On average, less than one percent of bony fish larvae metamorphose into adults, although certain species, such as sea horses, have evolved methods of parental care to enlarge this fraction, but the hammerhead shark may produce up to forty live young in a single birth.

The main prey of sharks range from molluscs and

1

2

3

crustaceans to large fish and marine mammals; and many are also efficient scavengers. The pig shark, or humantin, is a bottom-dwelling species which has grinding teeth in order to feed on tough sea urchins and shelled molluscs, but the numerous predatory sharks have different methods of feeding. The thresher shark, for example, batters its victims senseless with its body before feeding, while a curious beast known as the cookie-cutter shark has a tiny mouth with which it bites into a large fish or marine mammal as it swims by, allowing the combined force of the current and the velocity of its prey to swivel it round, thus removing a neat, circular portion of flesh. The majority of sharks, however, having bitten into their victims, thrash their bodies from side to side, so converting their teeth into a saw to cut the flesh. When one or more of the front teeth is lost or wears out, a new one will move forward to fill the gap from the row behind; almost all sharks have a lifetime's supply of teeth.

Surprisingly, perhaps, the largest sharks are peaceful, plankton-feeding creatures. At an average of eighteen-metres in length, the whale shark is the world's largest fish, reaching an average weight of some thirteen tonnes. The twelve-metre basking shark can process over 1,000 tonnes of water in an hour, passing it through its mouth and out through its gill

4

rakers, extracting the plankton en route. The basking shark is primarily a surface feeder, and is thought to shed its gill rakers in the winter when plankton is scarce, hibernating in the depths without feeding until favourable conditions return, when a new series of gill rakers develop. A third species of plankton-feeding shark was caught near Hawaii in 1976, and has been nicknamed 'megamouth'. It is thought to feed at greater depths, possibly luring small marine organisms by means of a luminescent membrane inside its mouth. Like whales and basking sharks it possesses several

Because more than fifty per cent of their body weight is blubber, causing them to float when dead, and also because of their slow-moving, surface-dwelling habits, the name right whale was coined by early whalers because these were the 'right' species to hunt. The black, or southern, right whale (1), which may weigh up to seventy tonnes, feeds mainly on the shrimp-like crustaceans known as krill (4) which abound in the southern oceans. Their heads are often covered with parasites, especially whale lice.

5

most hydrodynamic shape of any fish, their pectoral fins tucked away in grooves and their streamlined tails resembling the sickle moon. The sailfish, for example, can reach speeds of about 110 kilometres per hour – faster even than the cheetah, which holds the land speed record. Like sharks, these species swim with their mouths open in order to obtain sufficient oxygen to maintain such high speeds, and they too will drown if they stop swimming.

There is little shelter available in the open sea, especially in the light upper layers, but many animals are nevertheless able to avoid such ravenous predators. The tarpon, for example, is a large fish with shiny, almost mirror-like scales which, by reflecting the sunlight filtering through from above, render it almost

6

Toothed whales are highly predatory creatures, the ruthless, pack-hunting killer whales (**7**) having a reputation for playing with their victims rather like cats play with mice. Pilot whales (**3**) are not dissimilar to killer whales, being black with a white or pale grey breast plate, although their maximum length is only about eight-and-a-half metres. They are highly gregarious, with schools frequently numbering over 200 individuals. The white beluga whale (**5**) of the Arctic feeds mainly on prawns and medium-sized fish and embarks on a long migration southwards in search of food during the winter.

7

Great black-backed gulls (**2**) take advantage of the abundance of marine life at the ocean surface, swooping down to collect an unsuspecting fish in the beak without otherwise touching the water. Lesser frigatebirds (**6**), on the other hand, live by piratical means, chasing gulls, cormorants and boobies, pecking them until they drop their catch, and then recovering it in mid air.

rows of rudimentary teeth.

Despite the shark's reputation for aggression there are certainly more vicious and rapacious fish in the sea – notably the billfish. Many of these, such as barracudas, tunny and bluefish, form large packs which leave trails of blood and fragments of flesh behind them as they eat their way through shoals of lesser fish. The solitary marlin and swordfish lash out with their spear-like upper jaws as they pass through a dense shoal, consuming the dead and wounded on the return journey. The voracious barracuda may be more than three-metres long, but the largest bony fish is the swordfish – some individuals have reached a phenomenal six metres in length. All billfish can move at phenomenal speeds through the water, having the

invisible, whilst other fish, particularly those of the herring family, have blue-green backs and silvery undersides so as to blend in with the background and thus avoid the attentions of predators swimming either above or below them. Flying fish have greatly enlarged pectoral fins and can take to the air to escape predators, propelled by a violent thrust of the tail. Once clear of the water, these fins catch the sea breeze, enabling flying fish to soar for up to 400 metres before returning to the sea. Lower forms of life such as Venus's girdles, comb jellies and colonial salps avoid lesser predators by having transparent bodies through which the light passes almost as easily as through sea water.

Small, dark-bodied fish, especially plant-eating species of the open seas, form shoals which may number several million individuals. Each fish maintains an exact distance from its neighbour by means of its sensitive lateral line organs, which can detect pressure changes in the surrounding water, thus enabling them to change direction in unison. The shoal formation may serve to confuse predators so that they are unable to select a single target, or to deceive them into thinking that the whole shoal is in fact a huge fish, although it is not an adequate defense against such predators as the barracuda, which themselves form shoals. The slipstream effect may also serve to reduce the amount of energy used in swimming, much in the same way as gulls and geese fly in characteristic arrow formations.

Descending into the depths, such forms of camouflage become less essential under the cloak of permanent darkness. Yet some species of these lower realms seem to advertise their presence by producing light which is known as bioluminescence because it is produced by a living organism. Lantern fish, some squid, shrimps and hatchet fish, for example, are the marine equivalents of fireflies and glow-worms. Some utilise symbiotic luminous bacteria contained within special organs known as photophores to produce light, but many other species manufacture their own light-producing compounds called luciferins.

Bioluminescence serves a number of functions, one of which is to attract food. In anglerfish the dorsal fin is modified to form a sort of fishing rod which is suspended over its mouth and waved gently to lure victims within striking distance; in deep-water species, this decoy is decorated with a small luminous organ at the tip. The patterns of photophores may also help individuals of the same species to recognise one another in this twilight world, acting as a sort of sexual communication, but the ventral location of many of these light-producing organs has also led to a theory that they could serve as camouflage. At levels where light still filters down from above, any creature would be clearly visible to predators lurking below. Bioluminescence blurs the margins of its shape, and thus aids the fish or squid to remain undetected. Generally speaking bioluminescent fish living at higher levels have larger photophores and produce more light so as to camouflage themselves effectively in the higher levels of sunlight.

Bioluminescence is a common phenomenon in the mesopelagic realm, between about 300 and 2,000 metres; where large eyes are also a feature of many fish. Descending further into the true abyss, the darkness is complete, the waters cold, even at the equator, and the pressure may exceed 1,000 atmospheres in the deepest parts. Although these ocean depths were thought for many years to be lifeless, they are now recognised as the domain of more than 2,000 species of fish and invertebrates. Suddenly, bioluminescence is no longer a prominent feature of the inhabitants, and instead of having large, bulbous eyes geared to life in the twilight zone, many species are completely blind, their eyes having degenerated to mere pinpricks. Sight is useless here, so touch has developed to help creatures locate their prey and mates: long sensory feelers and spines make an appearance, such as those borne by the tripod fish, bristle mouth and blind lobster.

There are only two types of creature which can survive in the ocean depths: detritivores and predators. The detritivores are largely sedentary organisms, feeding on the gentle rain of dead organisms that drifts down from the productive sunlit levels. Most of them are filter-feeding or burrowing worms and molluscs. During the descent of the dead organisms, which may take more than three weeks for a small crustacean, the process of decomposition is already underway, and by the time the organism reaches the bottom, it may be only a skeleton, with very little nutrititive value. The

deep-sea food chain is thus geared to a starvation diet, so fish such as gulpers and swallowers have huge mouths and distensible stomachs capable of making the most of any food item that they might come across, even if it is twice their own normal body size.

There are also obvious problems in finding a mate in this realm of darkness, but some species of deep-sea anglerfish have solved the problem very neatly. The males are tiny creatures. A chance encounter with a large female is not to be passed up, so they bite into her flesh near the tail. Gradually their organs degenerate, their circulatory system merges with that of the female, and they become little more than repositories of sperm, suitably placed to fertilise any eggs that the female may produce during the course of her life. Sometimes a female may carry several of these 'parasitic' males around with her.

1

Typical ocean-going seabirds include two species of tube-nose: the waved albatross (1), noted for its endurance for long periods on the wing far from land, and the Cape gannet (2), which dives beneath the waves for its prey, wings held back like the fletches of an arrow.

2

3

The world's largest living creature, the blue whale (3) weighs up to 140 tonnes. Like most baleen whales, blue whales migrate to the poles in the spring to spend the summer where plankton is plentiful, returning fit and healthy to the tropics in the autumn to reproduce.

Most of these deep sea creatures are small compared with related species in the sunlit waters above, but some have gone to the opposite extreme. It is possible that the low temperatures and delay in reaching sexual maturity permit these animals to carry on growing for much longer than their counterparts near the surface. Medieval legends among seafaring folk revolve around such leviathans of the deep, and many a successful film about sea monsters has capitalised upon man's primeval fear of the unknown. Although for a long time such tales were regarded as pure imagination, towards the end of the nineteenth century verification came in the form of a giant squid washed up on a New Zealand beach, its tentacles measuring almost fifteen metres in

birds of the open sea. Most of these belong to the order *Procellariiformes*, or tubenoses, which contains about a hundred species, only a few of which, such as the giant petrel, do not depend directly on the sea for sustenance. The most impressive are undoubtedly the albatrosses, among which the wanderer has the largest wingspan of any bird: up to four metres in mature individuals. Other members of this group include petrels, shearwaters and fulmars, which are clumsy on land but masters of the sea breeze. They may spend nine months of the year of the wing, seeking land only to nest and rear their young, often on remote oceanic islands.

Many of these birds feed at night, timing their

Beluga whales (5), which frequent the Arctic Ocean, are pure white for camouflage against the pack ice, though at birth the calves are dark grey or brown. Due to their noisy and gregarious habits, these small-toothed whales have been nicknamed sea canaries.

The Feeding Grounds and Migration Routes of Northern Hemisphere Blue and Grey Whales

Blue Whale Feeding Ground

→ Migratory Route

Grey Whale Feeding Ground

→ Migratory Route

4

length and its body some three metres. Since then these monster cephalopods have become more familiar. We know, for example, that they rarely enter the light-ridden zones of the sea, can swim at over twenty kilometres per hour, usually measure between thirteen and eighteen metres across including their 'arms', and feed mainly on smaller cephalopods and fish. It is also suspected, although the scene has never been witnessed, that they enter into titanic battles with sperm whales, their sole natural enemy.

The picture so far seems to suggest that marine organisms are more or less confined to one particular ocean level, and that all movement of animals and their remains is downwards. In fact there exists a diurnal cycle of vertical migration of sea creatures in almost all the oceans of the world. It is a common phenomenon among the zooplankton and is probably associated with predator avoidance. As the sun goes down and the light intensity at the surface decreases, many small creatures that spend the day in the dark lower levels ascend to the surface, where they will spend the night feeding on phytoplankton before descending back into safety just before dawn. The daily migration of small predatory creatures is thought to be less related to changes in light intensity than to the ascent of their prey species. On the same principle, larger carnivores follow suit.

Other creatures which take advantage of this nightly influx of food to the surface layers of the sea are the

5

The grey whale (4) is found mainly in coastal waters and was once almost decimated due to excessive whaling activity. The eastern Pacific population was apparently completely annihilated by the beginning of this century, but a handful of grey whales that reappeared off California has been strictly protected ever since, and this population now numbers several thousand. It is not a large species, measuring less than fourteen metres at maturity and weighing in at only about twenty tonnes.

179

hunting activities to coincide with the period during which marine creatures are at their most vulnerable in the surface waters of the seas. Some species are sufficiently agile to be able to seize fish while on the wing or, like skimmers, dart across the surface with their beaks open to collect plankton. Others, such as most of the fourteen species of albatross, must actually settle on the surface to take their prey as they fly too fast to be able to pick up food on the wing. Yet others are adapted to plunging into the water, wings folded, to spear fish below the surface; the masters of this technique are undoubtedly the gannets, tropicbirds and brown pelicans. Lastly, some seabirds, such as penguins, greater shearwaters, cormorants and guillemots, are able to 'fly' underwater in pursuit of their prey.

Like the penguin and the albatross, seals are totally dependent upon the sea for their food, but must return to land to breed and moult. Yet they are well on the way to becoming true marine creatures. Their feet have become flippers and their bodies streamlined, and they have evolved methods of solving the problems ocean-going mammals experience. In order to sleep whilst at sea some have a thick layer of blubber which enables them to float upright like bottles, thus keeping their heads above water, whilst others doze beneath the surface, rising periodically to the surface to gulp air without waking. Whales, however, have gone one step further than the seal: having once conquered the land they have now returned to the sea, and have become so well-suited to their marine environment that they are unable to leave it.

1

2

Like all baleen whales, the grey whale (**2**) feeds on small fish and plankton which enter the mouth along with tonnes of water and are then squeezed against 150 pairs of thick baleen plates by the massive tongue. Killer whales (**3**) have no enemies except man, but even so they are not common, the population being limited by the availability of large fish, their main prey. When these are scarce, killer whales will resort to seizing seals and penguins, even approaching the shore to take an unwary elephant seal (**4**). Bottle-nosed dolphins (**6**) are the best-known of all cetaceans as they are often kept in captivity. In the wild they inhabit temperate and tropical seas, are non-migratory and live in schools composed of individuals of all ages.

The whales, or *Cetacea*, have a number of distinguishing features. Their tails are horizontal rather than vertical, their dorsal fins have no skeletal support, and their nostrils (the blowhole) are on tops of their heads. Their forelimbs are modified into flippers but their rear limbs are absent – even the pelvic girdle has disappeared, except in the Greenland right whale, which retains vestiges of this bone structure. The whales include the largest animal ever to have existed on this planet – the blue whale – which has been able to attain such a size because it is supported and protected by the surrounding water. On land a creature this big would have severe problems with the weight of the skeletal support and the musculature required to

move it, as well as with its temperature regulation. In effect, the blue whale has been freed from the constraints of gravity, and so is not required to spend vast amounts of energy in supporting and moving its enormous bulk.

In addition, whales display several distinctive physical and behavioural attributes which enable them to survive in the oceans. All whales have lungs positioned dorsally so as to make them more buoyant and, to enable them to spend long periods of time beneath the surface, about ninety per cent of each breath is converted from oxygen to carbon dioxide, as compared to only about twenty percent in man. When the air is released from the lungs it creates the typical spout – comprising mucus and water vapour – that was so useful to whalers in locating surfacing whales. Many cetaceans, especially migratory species, build up a thick layer of blubber for protection against the cold in polar regions, although their flippers are efficient heat exchangers, allowing the blood to cool if the mammal is too hot.

Perhaps the most advanced of all whales are the sixty-odd species of dolphins, which have been proved to be highly intelligent – probably more so than the most highly evolved anthropoid apes – and have a form of echo-location which is more sensitive than the most modern sonar equipment. In addition they have a complex social organisation, some living in schools of up to 1,000 individuals. They are capable of reaching speeds of up to thirty kilometres per hour,

oceans, and as yet none have become extinct, in spite of over-hunting. Nevertheless, there are threats more insidious than commercial exploitation that must be attended to before they and all other forms of marine life can be considered 'safe'. Contamination of the ocean with raw sewage, oil spills and radioactive wastes are threatening its myriad life-forms, while every year more than two million seabirds and 100,000 mammals die from ingesting discarded plastic containers or becoming trapped in abandoned fishing tackle. Although to date the oceans have proved equal to the task of absorbing the debris produced by mankind, the lesson is clear. No longer can we afford to treat the oceans as a dustbin.

When frigatebirds (5) make the effort to hunt for themselves their viciously hooked beaks are ideal for seizing young chicks or eggs from breeding seabird colonies. Gulls, on the other hand, prefer to pluck their victims from the surface of the sea (1).

3

4

5

and in doing so leave no wake, so smooth and streamlined are their bodies.

The Basques of northern Spain and southwest France are thought to have been the pioneers of the whaling industry, practising this specialised skill at least as long ago as the twelfth century, if not earlier. By the sixteenth century, having depleted the whale stocks of the Bay of Biscay, they were roaming as far afield as Labrador and Newfoundland in search of their prey. The Eskimos too have a long tradition of whale hunting, and in some instances have been completely dependent upon these creatures for their existence. In both cases the hunt was a matter of a few men pitting their skill against this marine mammal, and as such had little effect on whale numbers. However, when man began to apply his technological prowess to the problem, the whales had no defences, and soon numbers began to fall.

By 1935 it was obvious that whales were disappearing rapidly and a law was passed protecting all right whales from commercial harvest. In 1977 even aboriginal hunting of whales was forbidden by the International Whaling Commission. For several years now moves have been afoot to protect all whale species, although several countries, notably Japan, still refuse to halt their whaling activities completely, even though threatened with economic sanctions by more concerned nations.

Whales are the ultimate symbol of life in the

6

· CHAPTER 8 ·

COASTLANDS

Taken as a whole, coastlines form some of the most diverse and productive ecosystems in the world, ranging from coral reefs and mangrove swamps on tropical shores to the more ubiquitous sandy or rocky beaches and mudflats elsewhere. The myriad life forms they support encompass both the kaleidoscopic creatures that are dependent upon the tropical reef formations, and the dull grey and brown animals that bury themselves in the estuarine muds. Almost all animal phyla are represented in the coastal systems of the world. Some, such as seals and manatees, have returned to the ocean having conquered the land; others, including a vast array of fish and invertebrates, have never escaped from the confines of the marine environment. But the most characteristic animals of the coasts are those amphibious creatures poised to make the prodigious evolutionary leap from the sea to a terrestrial habitat.

Right: a spotted coral trout, (top) fanworms and (above) yellow-horned poppies.

The earth's coastlines, including both mainland and islands, have been estimated to extend over more than one and a half million kilometres. Without a doubt they are among the most diverse ecosystems in the world, varying from the wall of ice which towers over the ragged fringe of the Antarctic peninsula to the pure white sands, aquamarine lagoons and gently waving palms of a tropical island bay; from malodorous muds and dripping prop roots in the deep shade of a mangrove swamp to vivid undersea meadows; from deep coastal fjords and rias to sweeping expanses of sand and shingle and from the swirling colours and myriad life forms of a coral reef to sheer cliffs splashed with the guano of vociferous seabird colonies.

The shape of a coast is influenced by many factors. In the long term it is affected by changing sea levels. Since the ice ages the sea has been rising, by about thirty centimetres per year, as is indicated by the presence of submerged ancient forests that are sometimes exposed at very low tides. Furthermore, this meeting place of land and sea is not a static environment, but is constantly changing. Where exposed to the full force of the sea, the land is in retreat; where it is sheltered, maybe only a few kilometres along the coast at the mouth of a large river, terrestrial habitat is encroaching little by little into the marine environment. The physical nature of the land also affects the configuration of the coast. Where the rock

is hard it resists the erosive force of the sea, often forming sheer cliffs; where soft, it is eaten away more rapidly.

Tides are the single most important factor affecting the world's coastlines. They are created by the gravitational pull of the moon and the sun. Although the moon is smaller, its influence is much greater than that of the sun because it is so much closer to the earth. The gravitational pull that it exerts distorts the ocean, causing it to bulge slightly at two diametrically opposed points. As the moon moves around the earth, and the earth spins on its axis, the water bulge moves across the face of the planet, creating higher sea levels in each area it traverses, which cause high tides at the coast.

Seahorses (1) are unique among fish in having their heads set at right angles to their armour-plated bodies. The tail fin is absent and the lower part of the body is prehensile to facilitate anchorage. The Portuguese man o' war (5) is a colonial siphonophore with a thirty-centimetre gas-filled float.

Most parts of the world experience two such high tides in every twenty-four-hour period, separated by periods of low water as the bulge moves onwards.

The approximately monthly cycle of the moon confers an additional pattern on the diurnal tidal sequence. About every two weeks, in conjunction with the new and full moons, the moon and sun are aligned with respect to the earth, and the gravitational forces they exert are therefore stronger, producing what are known as spring tides, which ride much further up and down the shore. Alternate weeks of the month, during the first or last quarter of the moon, see the sun at right angles to the moon, thus counteracting its pull. At these times neap tides, which have a lesser tidal range, are experienced. Furthermore, the spring and autumn equinoxes are responsible for exceptionally high tides, which can cause flooding far inland if they should happen to coincide with storms. Conversely, the solstices are times of very low water.

Not all shores experience identical tidal ranges, however, since the gravitational pull is much less evident in small, enclosed bodies of water, such as the Baltic and Mediterranean seas. In the Mediterranean the tidal range is almost negligible, the difference between high and low tide being less than fifty centimetres. On the other hand, shores facing a large expanse of open ocean can suffer from huge tidal ranges. The highest recorded differences under normal

5

Harbour, or common, seals (**2**) are found in temperate waters between the Arctic and the Mediterranean on both sides of the Atlantic. They have a predilection for flatfish. Black-winged stilts (**4**) are cosmopolitan birds found from Europe to Australia. Their distinguishing physical characteristic are their long red-pink legs, which allow the bird to exploit deeper water than other waders. White-bellied sea eagles (**3**), their range stretching from India east to China and south to Australasia, rarely stray far from their coastal territories.

conditions are experienced in the Bay of Fundy, Nova Scotia, where a tidal range of up to eighteen metres is common.

The littoral zone of the shore corresponds to that area lying between the upper limit of the spring tides – marked by the strandline – and their lowest level, below which the shore is only rarely exposed. It is this region which is most affected by the constant ebbing and flowing of the tides, creating a harsh environment for the organisms which live there. Regardless of the physical characteristics of the shore, all littoral animals and plants must be able to cope with the twice-daily exposure to the atmosphere, as well as the battering force of the waves, which reach pressures of up to

1

Thrift (**2**) may form vast sheets on coastal cliffs and saltmarshes throughout Europe, sea bindweed (**7**) is a creeping, fleshy-leaved plant of Atlantic European sand dunes, and the bizarre ice-plant (**3**), native to South Africa, has become naturalised on Mediterranean and Californian shores.

2

thirty tonnes per square metre on exposed Atlantic coasts. But despite these conditions, the littoral zone teems with animals and plants taking advantage of the abundant supply of light, oxygen, water and minerals in such places.

Aside from variations in tidal range, the extent of the littoral zone is also dependent upon the gradient of the shore. On steep coastlines there may be only a few metres between high and low tides; but in some parts of the world, where the slope of the shore is negligible, the sea may be almost out of sight from the strandline at low tide. Nevertheless, all shores display a zonation of intertidal organisms dependent mainly on the ability of these creatures to withstand exposure to the atmosphere at low tide, but also affected by various interspecific relationships, such as competition and predation.

At the lowest levels, which are uncovered for only short periods before the tide reverses its flow and begins to advance up the shore once more, the predominant life forms are those adapted to a marine way of life. At the opposite extreme, the strandline is occupied only by a few creatures that appear to be making the transition between a marine and a terrestrial existence. True land animals, such as scavenging bears and hyenas, known as strand wolves in coastal areas,

3

are also found there, having learned to exploit this transient niche. Of course, when the tide is full the shore zonation is less obvious since true marine animals are able to take advantage of the enveloping waters. On the whole, animals and plants are able to exist much higher up the shore in temperate parts of the world than in tropical ones since there is less risk of desiccation at low tide in cooler climates.

Because the opportunities for concealment are few on a rocky shore, the zonation of intertidal organisms is very conspicuous in such places. There are three main zones in evidence on the temperate rocky coasts of the world. At the highest level, which may be reached by the sea only during the spring tides or

Roseate spoonbills (**1**,**6**) are found throughout the New World tropics, with two small North American populations in Texas and the Everglades. There are about 1,600 species of brittle-star (**4**) in the world, found both in coastal habitats and in the ocean depths. At low tide, mangrove swamps usually teem with crabs, especially members of the tropical family Grapsidae (**5**).

during stormy weather, the rocks are daubed with lichens tolerant of salt-spray and a slimy film of blue-green algae. A careful examination of the crevices and fissures at this level will reveal a type of periwinkle which is supremely adapted to life without the sea. Not only is it able to breathe through lung-like gills, which require only occasional moistening to be able to extract oxygen from the atmosphere, but it also hatches its young within its shell instead of laying eggs, which would quickly dehydrate in these conditions.

Conditions are most stressful in the middle section of the shore since it is alternately exposed to the air and inundated by the tide twice a day. The most obvious feature of this zone is the broad band of brown algae, more usually known as seaweed. Seaweeds can in fact be red, brown or green, depending on which pigment is predominant. All seaweeds also contain chlorophyll and photosynthesise in exactly the same way as land plants, so they need to be situated in shallow waters where sufficient light is available for them to do so. But

1

2

3

here their resemblance to land plants ends. The complex internal organisation of a land plant is quite absent in seaweeds since they have no need of specialised structures to support them, to transport nutrients within them or to enable them to reproduce. The sea carries out all these functions for them. The only obvious division of labour catered for in the structure of a typical seaweed is the distinction between the lamina and the holdfast, which provides the plant with anchorage.

The brown algae typical of the middle zone of a rocky shore are wracks, often bearing small bladders filled with air or a gelatinous substance to buoy them up towards the light when the tide is in. Also common in this mid-tidal region are several species of green algae, such as sea lettuce. Despite the abundance of seaweeds, there are in fact very few species that actually eat them. However, most seaweeds produce a coating of mucus that prevents them from drying out whilst the tide is low. This is taken advantage of as cover by such creatures as tiny, soft-bodied crabs and periwinkles, which are not as tolerant of exposure to the air as their relatives further up the shore.

Another feature of this middle region of the shore is the strip of white acorn barnacles that winds across the rocks. When the planktonic larval forms of this crustacean are mature, they locate a suitable site for

Glanville fritillary butterflies (**1**,**5**) are found from Western Europe to the Amur River bordering Manchuria, in flowery meadows up to an altitude of 1,800 metres. In Britain, however, because they are on the northern and western limit of their range, they have a tenuous hold only on the most southerly cliff-top grassland of the Isle of Wight, where their caterpillars feed on ribwort plantain and bask in the sun to raise their body temperatures. Painted storks (**2**) are found widely distributed across the freshwater wetlands and coastal swamps of Southeast Asia, nesting in colonies in the tops of mangroves and other wetland trees.

settlement on the shore and attach themselves to the rock by means of suckers on their antennae. Each barnacle immediately starts to secrete protective limestone plates around itself, which form a conical turret. It will not move again during its lifetime, which may span three to five years depending on its position on the shore. Barnacles are only able to feed when covered with water, and do so by drawing back the plates in the top of their carapace and rhythmically flexing their legs, which have been modified into bristly cirri, thus filtering out small particles of organic matter and plankton from the surrounding water. Those

which are exposed more frequently to the air can feed less often, grow more slowly and thus live longer.

There may be as many as 30,000 acorn barnacles per square metre, so tightly are they packed, but more often than not they form a mixed belt with slate-blue mussels. Each of these bivalves is attached to the rock by a few strong but flexible byssus threads, which allow them some 'give' in the face of powerful waves whilst preventing them from being torn loose altogether. The position of the barnacle and mussel zone on the shore depends not only on their ability to withstand exposure to the air, but also on the actions of their predators. Both are preyed upon by dog whelks – gastropod molluscs armed with horny-toothed radula with which they drill holes in the shells of their victims in order to extract the contents. Mussels are also a favourite food of various species of starfish which are able to lever the two halves of the shell open and extrude their stomachs to digest the unprotected tissues of the mussel within. These predators, although tolerant of exposure to the air, cannot feed whilst the tide is out, so the higher up the shore the mussels and barnacles can survive, the less time they will spend in the water and the fewer individuals will succumb to voracious dog whelks and starfish.

Another successful middle zone mollusc is the limpet. These clamp their conical shells tightly to the rock both when the seas are strong, so that they don't get dislodged, and also when the tide is out in order to preserve moisture between shell and substratum and so prevent desiccation. The foot muscle which achieves this is so powerful that it is almost impossible to remove a limpet from its chosen perch without the aid of a sharp knife to lever under the edge of the thick shell. No doubt this is a useful defence against the attentions of seagulls and other shore-feeding creatures. Sea anemones may also be encountered on the middle shore, although they favour overhanging ledges or crevices where they are less exposed to sun and wind. Submerged sea anemones feed by extending their tentacles, each bearing a battery of stinging cells called nematocysts, and paralysing small shrimps or young fish which come within reach. At low tide, however, anemones retract their delicate tentacles and become little more than amorphous blobs of jelly hanging from the rocks.

As might be expected, the lowest zone of the shore is the richest both in abundance and diversity of organisms since it is closest to a true marine environment. It is exposed only during the ebbing of the spring tides. The brown seaweeds are still present at this level, only here they are represented by dense, offshore forests of kelp with enormous holdfasts and laminae up to a hundred metres long, often festooned

The purple gallinule (**3**) of the New World tropics and the subtropical coastal wetlands of the southeastern United States is closely related to rails, coots and moorhens. Its name is derived from the Latin meaning 'little hen'.

with colonies of lace-like bryozoans and delicate hydroids. Red seaweeds are also present in a variety of shapes and sizes including cushions, ribbons and slender hair-like filaments. Mobile creatures abound here, especially predators such as long-spined sea urchins and spider crabs, as well as swimming velvet crabs and many fish species – including rock eels, gobies and rock bass – which can retreat to the safety of the oceans on the rare occasions when this part of the shore is exposed.

Sometimes these lower-shore animals are found higher up the beach in rockpools, stranded by the retreating tide. These miniature seas provide a fascinating glimpse of life at the lowest levels of the shore. Some sedentary organisms, such as seaweeds, mussels, barnacles and fanworms, are permanent inhabitants of tidal rockpools, but starfish, hermit crabs, shrimps, sea hares and an occasional octopus or goby are able to escape the pools with the next tide.

Life in these tidal pools is precarious, increasingly so the higher up the shore they lie. Hot weather raises the rate of evaporation from rockpools, which consequently become more saline and may cause its inhabitants to shrivel as a result of water loss from their bodies. Conversely, prolonged rain at low tide will serve to dilute the waters, causing the occupants to absorb water and become bloated. In either case, extreme variation in salinity will result in death. The availability of oxygen is also reduced in warm waters, and creatures stranded within overheated rockpools will, in effect, suffocate.

Behind almost every rocky shore lies a line of cliffs which form the effective boundary between the sea and the land. Often the highest tides pound the cliff base, sending spray far up its face and creating conditions which are suitable for the growth of halophytic plants. Sea thrift, sea spleenwort, rock samphire and sea

1

2

Gulls are considered to be among the most successful birds in the world today. There are some forty-five species of gull, all of which are strong fliers and swimmers and opportunistic feeders, distributed from the Arctic to the Antarctic and from sea level to the high mountains.

3

4

spurrey adorn the ledges of many cliffs in temperate regions, along with many brightly coloured lichens. Northern shores support such species as roseroot, moss campion and lovage, while more southerly cliffs are swathed in kidney vetch, sea beet, wild carrot and golden samphire. The short maritime grassland at the top of the cliff is often rich in herbs, although wind and sea spray reduce them to dwarfed replicas of the same species growing inland. Spring squill is a characteristic European cliff-top species, as are sea campion, sea milkwort and sea fern grass. Trees are usually absent, but those that do manage to grow are heavily wind-pruned, giving them the appearance of leaning away from the shore.

Such sea cliffs are usually frequented by large, noisy colonies of breeding and roosting seabirds. On vertical cliffs in Europe, guillemots nest on exposed rock ledges. Their eggs are rounded at one end and almost pointed at the other so that they roll only in circles and not onto the rocks below. Kittiwakes build tattered nests of seaweed and grass attached to impossibly sheer rock faces. Razorbills and black guillemots favour fissures and less exposed sites behind boulders; manx shearwaters and puffins burrow into the friable soil on cliff-tops and eider ducks frequently build their nests on rocky shores close behind the strandline. Other common cliff-nesting seabirds include shags and cormorants, little auks and fulmars, and more than seventy per cent of the world's population of gannets nests on the cliffs and offshore stacks of the British Isles.

5

These large concentrations of seabirds rarely frequent the rocky shore to feed, however, but obtain most of their sustenance directly from the sea. Other birds do come to rocky shores for this purpose, notably oystercatchers, which prise open mussels with their powerful red-orange beaks. Purple sandpipers eat winkles, crabs and dog whelks, and rock pipits both feed and breed on the upper margins of the shore. Gulls and terns sometimes scavenge along the strandline, and are among the few creatures able to prise limpets from the rocks. Birds of prey and scavengers – particularly sea eagles, peregrine falcons, ravens and choughs – will often take advantage of the abundance of eggs and nestlings present in seabird colonies, themselves nesting and rearing their young nearby.

6

8

Herring gulls (**1**) breed on sea cliffs, dunes, shingle and marine islands across most of the northern hemisphere, feeding both in coastal and inland regions, while the cliff-nesting western gull (**5**) is a species of the Pacific shores of North America, and the laughing gull (**2**) is a denizen of the southeastern coastline of the United States.

7

Cormorants are expert fishers with stiff tails that act as underwater rudders and powerful webbed feet set towards the rear of the body. The absence of oil glands means that their feathers become waterlogged easily, and they are obliged to spend long periods basking in the sun with wings outstretched. Common cormorants (**4**) breed on all types of coast, but the smaller shag (**7**) prefers inaccessible cliff-faces. Cape gannets (**3**) breed in huge colonies on South Africa's offshore islands, while puffins (**6**), the smallest breeding auks south of the Arctic, construct burrows in cliff-top turf in which to rear their young. Guillemots (**8**) breed in close-packed colonies on rocky ledges in the northern hemisphere, their single eggs being pear-shaped so as to roll in circles and not onto the rocks below.

The only mammals that are truly adapted to take advantage of the myriad life forms which live on rocky shores are the sea otters of the Pacific shores of North America and eastern Asia. They frequent offshore kelp forests, wrapping themselves in fronds of this seaweed when they sleep in order to reduce the disturbing effect of wave motion and to prevent their drifting out into the ocean. The sea otter forages among the kelp, feeding on other denizens of the kelp forest; but it also has a predilection for shellfish, which it smashes open against a rock carried on its chest whilst it floats on its back.

As exposed cliffs are eroded by the sea and retreat inland, the waves carry away the resulting rock debris to sheltered places further along the coastline where less violent waves, instead of eroding, deposit their load of sand or gravel on the shore. In this way the erosion of rocky parts of the coast contributes to the formation of sand and shingle shores elsewhere. Shingle beaches are usually found where the nearby cliffs are either very hard or contain hard chunks within a soft matrix, such as the flint nodules in the chalk coast of

The Distribution of
Sea Otter Populations
along the Northern
and Eastern Pacific
Coastal Rim

Sea otters, the largest mustelids in the world, are today found mainly off the Californian and west Alaskan coasts (**1**,**2**), although once they were distributed across all shores of the North Pacific, their decline being due to the pressures of the fur trade. At night they wrap themselves in strands of kelp (**3**) so as not to drift out to sea, and they are renowned for their use of rocks to open tough bivalves (**5**). The frilled float of the Portuguese man o'war (**6**) can be a liability – as well as propelling the creature to new feeding grounds, if caught by a strong wind it can carry the creature to its death, stranded on an Atlantic beach. The Southeast Asian fishing cat (**8**) has adapted to life among the mangroves, feeding on the fish that teem in the shallow tidal waters. Coastal birds can contrast as sharply as the huge painted storks (**9**) of Southeast Asia and the cosmopolitan black-headed gull (**10**).

southeast England. The pieces of shattered rock gradually become smooth and rounded due to the constant action of the waves, which roll them against one another, so that the more time they spend in the sea the smaller they become. To qualify as shingle rather than sand, the pebbles must measure more than five millimetres in diameter, though they can be up to twenty centimetres across.

Shingle beaches display a gradation of pebbles, from the smallest nearest the water to the largest at the top of the beach, hurled there by the force of storm waves. On the whole, shingle beaches are inhospitable places for intertidal organisms. At the water's edge the pebbles are constantly churning in the backwash of the waves, in an action similar to that produced by a cement mixer. Unlike the solid, unchanging surface of the rocky shore, the mobile shingle provides no footholds for sedentary animals or plants. Sandhoppers shelter under piles of rotting seaweed at the high-tide mark during the day and, if stranded as the tide retreats, blennies may burrow into the shingle on the lower beach where there is still sufficient water to survive until the shore is once again covered by the sea.

However, at the top of the beach the build-up of organic material on the strandline allows the seeds of annual species such as sea beet and oraches to germinate as, at this level, the shore is only rarely disturbed by the tide. Above this level, out of reach of all but the most violent storm-driven seas, the shingle is more stable, and in it a wide range of attractive shore plants can be found, their deep roots reaching down between the pebbles to reach moisture and small amounts of soil that have accumulated over many years. Yellow-horned poppy, sea sandwort, oyster plant, sea pea and sea kale are all perennial species which grow on exposed shingle banks above the strandline in Europe.

Birds which take advantage of undisturbed stretches of shingle for breeding include ringed and Kentish plovers, the eggs and chicks of which are well-camouflaged to blend in with the pebbles. Several species of tern, including the Sandwich, roseate, arctic and little terns, frequently nest on offshore shingle banks. Oystercatchers and common and black-headed gulls often adopt the same strategy. Turnstones are among the few species that feed along the shingle, flipping over pebbles and piles of seaweed near the strandline, looking for sandhoppers.

Sandy shores, too, appear barren at first glance, but are in fact less hostile than they seem. Although there are no firm footholds in the shifting sands of the intertidal zone, many sedentary creatures are able to survive here by burrowing beneath the surface, out of reach of the churning action of the waves.

6

Rock samphire (4), a fleshy-leaved member of the carrot family, grows on coastal cliffs and shingle in western Europe, the attractive hare's tail grass (11) is native to the Mediterranean region, particularly among sand dunes, and the saw-sedge *Cladium jamaicensis* (7) prefers waterlogged, brackish habitats in the Everglades.

7

8

9

10

The lowest levels of the food chain consist of the particles of organic debris brought in by the waves, ninety-five per cent of which are rapidly consumed by surface-dwelling bacteria. Any larger animals which are to survive here must therefore aim to trap these particles before they come to rest. Some, such as sandmason worms, construct a tube of sand grains and mucus to raise their delicate network of tentacles above the surface. Others – especially bivalved molluscs such as razor shells, fan mussels, Venus shells, American geoducks and cockles and a few species of gastropod, including the pelican's foot and tower shells – bury themselves deep in the sand but extend siphons to the surface which draw a constant current of water down

into their burrows. The masked crab also adopts this approach, clasping its antennae together to form a tube which fulfils a function similar to that of the siphons.

As is the case on rocky shores, the animals of the sands are stratified according to their ability to survive while the tide is low, though the zonation is less evident. The upper reaches of the shore support many creatures which may be in the process of becoming independent of the sea, such as kelp flies and the ubiquitous sandhoppers, which thrive in the strandline debris. Sandhoppers, which may be present in densities of 25,000 per square metre, are small crustaceans with a hard carapace which enables them to withstand

11

desiccation and powerful legs to help them avoid predators. The ghost crabs which haunt the higher sandy shores of the American Atlantic coast may also be on their way to becoming truly terrestrial. Although they still have gills which need to be kept moist, they carry a supply of sea water around in a cavity surrounding these gills, returning to the sea for a fresh supply when the oxygen becomes depleted.

The true intertidal zone of a sandy shore is the almost exclusive domain of burrowing creatures, mostly bivalves and worms. Parchment worms and acorn worms construct U-shaped burrows through which they draw a constant supply of sea water that bears food particles to them when the tide is full. The parchment worm secretes a mucus net, which it proceeds to consume when the net is full of trapped organic debris. Lower down the shore, the colonial honeycomb worm creates small reefs of sand grains and shells, up to sixty centimetres across, on small rocks or half-buried timbers that protrude from the sand.

The lowest levels are inhabited by creatures which cannot withstand prolonged exposure to the air, including burrowing sea anemones, sea urchins and starfish, segmented worms, burrowing squid, shrimps, swimming crabs, sand eels, flatfish, skates and rays. Sea urchins have reduced spines, most of which point backwards, but they can be rotated to facilitate

tunnelling. Some, such as sand dollars, are flattened in a dorsoventral plane and live close to the surface, whilst others burrow to a greater depth, for example the more rounded heart urchins. In the majority of cases these urchins ingest sand and feed on the organic detritus and algae which cling to the grains. The burrowing starfish that live within this sandy environment have no suction cups on their tube-feet as they do not need to cling onto the substratum to withstand the force of the waves. Unlike those of the rocky shores, these predatory starfish eat small snails and crustaceans whole, regurgitating the shells later.

The dozen or so species of clownfish (**2**,**6**,**10**) enter into symbiotic relationships with the plantlike coelenterates known as sea anemones. These small damselfish from the Indo-Pacific Ocean are usually boldly marked in bright colours and spend their entire lives within the shelter of the stinging arms of the anemones, sometimes even retreating into the polyp's mouth if danger threatens. It is not certain why these fish are immune to the anemone's nematocysts, but it is possible that they have a protective mucous coating. The benefits to the anemone are also unknown, though some authorities suggest that the fish aerate it with their fins.

1

2

3

4

Coastal decapods include the scavenging hermit crabs, which occupy empty gastropod shells (**5**) and may carry a cloak anemone around for camouflage and protection (**7**), as well as those which have evolved the physical apparatus to spend a large proportion of their time on the shore, such as the fiddler crabs (**9**), the males of which have one outsized claw for signalling and ritual combat, and other occupants of the Southeast Asian mangrove swamps (**4**).

Many of the carnivorous species of the lower shore hunt by ambush, burying themselves in the sand until only their eyes protrude and launching themselves at their prey when it ventures within range. Torpedo rays in temperate waters and the toadfish of tropical sands are examples of such hunters, as is the rotund pomegranate crab. Most of these mobile species retreat with the ebbing tide, but others are sometimes seen far up the beach. Mole crabs advance with the tide and filter water-borne debris from the backwash of the waves, moving forward at intervals to stay within the surf where the turbulence favours such an existence. On African sandy beaches, plough snails erect their single ploughshare-shaped feet in order to catch the waves that will carry them up the beach, where they quickly locate beached jellyfish or other stranded marine creatures and gather round to feed. When the tide begins to retreat, these snails lift their feet into the surf once again in order not to be stranded themselves.

Other animals are not permanent inhabitants of the sandy shore, but may utilise it for breeding purposes. The Californian grunion is a small fish which spawns in the sand. Some internal clock, probably influenced by the lunar cycle, tells them when it is time to come ashore. They always appear just after the full moon when the tides are at their highest. Then females lay large quantities of eggs in the sand, which are rapidly fertilised by the males, and each adult then catches a

5

unable to right themselves. The larvae of both grunion and horseshoe crab develop rapidly so that by the time the next spring tide reaches them they are ready to be washed out into the ocean to join the zooplankton.

However, other species have learnt to take advantage of the precise swarming time of the horseshoe crab. May in Delaware Bay, between New York and Washington, sees up to a million shore birds lined up for dinner. Most of these are on their way north to breed in the Arctic tundra, having come from their wintering grounds in Central and South America. Delaware Bay, like other areas of mudflats on both

6

Shallow tropical seas host fireworms (1), members of the 8,000 species which make up the class of segmented polychaete worms, as well as brittle stars (3), distinguished from true starfish by their slender, flexible arms, which lack tube-feet, and their disc-like bodies.

7

8

9

wave back to the sea before the tide starts to ebb.

Horseshoe crabs display a similar sense of timing. These strange crab-like creatures are actually related to an extinct family of water scorpions and have changed little since Devonian times, some 360 million years ago. Today there are four species, three of which frequent mudflats and mangrove swamps in southeast Asia, but the fourth spawns on the sandy beaches of the North American Atlantic coast. During May and June they catch a spring tide, the males clustering around the larger females, to deposit their fertilised eggs in the sands on the highest part of the beach. They too return to the sea with the ebbing tide, save a few which have been tipped onto their backs and are

10

The cool, temperate North Sea cannot rival the colour and diversity of the tropics, but nevertheless supports large, attractive species such as the dahlia anemone (8).

1

Estuarine crocodiles (**5**), the largest living crocodilians, and Indo-Malayan water monitors (**3**) are both typical predators of mangrove swamps and have even been seen at sea. Green turtles (**1**) undertake long breeding migrations, those individuals frequenting the Brazilian coast covering some 4,500 kilometres to breed on Ascension island.

3

4

Pacific and Atlantic coasts, is one of the essential staging posts where birds must refuel to be able to complete the journey. Millions of small, turquoise or green-brown horseshoe crab eggs, about one millimetre in diameter, provide the basis of a three-week-long feeding orgy for, among others, sanderlings, red knot, ruddy turnstones, sandpipers, semi-palmated sand plovers, dunlin, red phalaropes and American avocets. The horseshoe crabs are able to sustain such losses as each female lays about 80,000 eggs, and there may be as many as 10,000 females per kilometre of shore. What is less certain, however, is whether the migratory birds of Delaware Bay, faced with a round trip of 12,000 to 25,000 kilometres per year, would be able to survive without the horseshoe crabs.

The other main group of marine creature using sandy beaches as a nursery are the turtles. There are six species of sea turtle in the world, divided into two families. The hard-shelled turtles – green, loggerhead, hawksbill, Kemp's Ridley and olive Ridley – resemble tortoises in that they are protected by a horny carapace; but the leathery or leatherback turtles, as their name suggests, bear only a thick leathery skin on their backs, raised into seven longitudinal ridges. They are the largest of all marine turtles, often measuring two metres in length and weighing up to 600 kilogrammes. All species of turtle live in tropical and semitropical seas, but the loggerhead's range also extends into

5

6

Female horseshoe crabs (**2**), which may be up to sixty centimetres in diameter, come ashore at certain times of year to lay up to 300 eggs (**6**) in the sand. These are then fertilised by the attendant males (**4**).

temperate regions. Although these creatures returned to the sea millions of years ago, they are still linked to the land in that their eggs require air to develop.

In all species the males never leave the water, but the females laboriously drag themselves up the beach, far beyond the level of the highest tides, to scrape out shallow holes with their flippers in which to lay their eggs. The female turtles regularly return to the same beaches for this purpose, but of thousands of apparently suitable sites only a few are used. The Ridley turtles, most of which do not grow to more than sixty centimetres in length, breed in vast numbers on the coasts of Mexico and Costa Rica; at times there may be

winds may reverse the process, propelling the sand dunes ever seawards.

At the base of the dunes that lie nearest to the sea, organic debris usually accumulates, enabling such species as prickly saltwort, sea rocket and common orache to grow. Above this, out of reach of all but spring tides and storm-driven waves, sand couch grass and lyme grass become established – the very existence of which helps to stabilise the dunes by trapping sand grains around their roots and leaves. They pave the way for marram grass, a typical component of dune vegetation. Its long, branching root systems further stabilise the dune face. Marram grass is highly

Herons are typical birds of both inland and coastal waters, wading through the shallows in search of fish and amphibians, which they spear with their swordlike beaks. The great blue heron (7) is a North American species, while the similar grey heron (8) is found in Eurasia and Africa.

7

more than 100,000 on a single beach. The leatherback is thought to breed in only two places in the world: on the east coast of a few Malaysian islands and along the shores of Surinam in South America. Rather than coming ashore in large numbers they emerge from the sea a few at a time at night, the same female often returning on several consecutive nights to lay tens rather than hundreds of eggs.

The young turtles must make their way down to the sea within a few hours of hatching. The main problem associated with the predictable use of the same beaches every year is that scores of predators – from crocodiles to seagulls – are waiting to take advantage of the feast. It is not surprising therefore that the proportion of young turtles to reach the sea in safety is very low indeed.

The action of winds and waves are constantly driving the sands of such beaches inland, where they often form parallel ridges of dunes. Although the dunes themselves do not actually move inland, the action of the wind carries individual grains over the crest so that the dune is gradually rolling over like a solid ocean wave. In other parts of the world, prevailing offshore

8

xerophytic, growing in a desert-like habitat where water is scarce. Its leaves are tightly rolled to avoid unnecessary water loss and its stomata are hidden within.

The presence of marram grass enables other dune species to invade, including such attractive plants as sea holly, sand sedge, sea daffodil, restharrow and sea bindweed. Inbetween these plants, lichens and mosses colonise the bare sand, together with sea sandwort and sea milkwort, so that a bare dune eventually becomes completely covered with vegetation. The low-lying areas between the rows of dunes are called dune slacks. Water frequently collects here, encouraging the establishment of species such as marsh helleborine, wintergreen and creeping willow. Around the Mediterranean, mature dunes may be colonised by umbrella pines, whilst dunes on tropical coasts are frequently stabilised by screw pines, with long, looping prop roots which help to anchor the shifting sands, as well as coconut and date palms.

1

2

3

Oystercatchers (**4**) are a remarkably homogeneous group, all species having black and white plumage, pink legs and feet and powerful, flattened orange beaks designed to open mussels and prise limpets from the rocks. Outside the breeding season they gather in huge flocks, dispersing into pairs in the spring to rear their young on shingle spits and rocky beaches. Collared pratincoles (**1**) are essentially African and Mediterranean birds, more attracted to freshwater than coastal regions; they are noisy, gregarious and capable of amazing aerial acrobatics *en masse*. Black-winged stilts (**2**), birds of brackish waters and saltmarshes, are related to avocets, and like them they scythe their slender bills from side to side through the bottom sediments in search of small aquatic invertebrates.

Many sandy shores are bordered by a bed of undersea meadows in the shallow waters beyond the littoral zone, just as the top of the beach is often fringed with dune grasslands. Marine meadows are found in both temperate and tropical waters, particularly off the coasts of Europe, North America, Asia and Australasia. They are the soft substratum equivalent of the dark kelp forests which predominate beyond the low water mark of rocky shores, growing in sheltered conditions at depths of between about six and fifty metres. But the plants which make up an undersea meadow differ from the rocky-shore seaweeds in that they are true vascular plants which, like whales, seals and turtles, have returned to the sea after conquering the land. There are more than thirty species of these 'sea grasses', so called because of their linear green leaves, which may be up to a metre long and less than one centimetre wide, although they are more closely related to pondweeds than to terrestrial grasses. Their long, branching root systems provide good anchorage in the soft sands and muds.

Marine meadows provide a home for many creatures which cannot survive without some form of shelter. Some small, colonial organisms, including hydroids, bryozoans, tube worms and red and green algae, exist epiphytically on these sea grasses. For other species, their leaves and seeds represent a source of food, not least for wintering ducks and geese in temperate zones, and for sea turtles, dugongs and manatees in warmer waters. Scallops shelter among the roots; seahorses, like living chessmen, curl their tails around leaves for anchorage, and long, slender pipefish are almost invisible as they hang motionless and vertical among the grasses. All in all, the teeming inhabitants of these undersea meadows constitute a complex food chain; one square metre of turtle grass off the tropical Atlantic coast has been estimated to support some 30,000 individual animals.

The muds and silts brought down to the coast as waterborne river sediments, and those carried ashore by the waves, accumulate along sheltered coasts, especially in estuaries. These particles are even finer than the grains which provide the matrix of the sandy shore and, as such, become packed more closely together, inhibiting the circulation of water. Thus burrowing creatures which breathe predominantly through their skins, such as some urchins and starfish, are absent from muddy shores. But these sediments are rich in organic materials and are often coated with

diatoms and blue-green algae on the surface, providing vast food reserves for creatures which obtain their oxygen by other means.

As on sandy shores, bivalves are very common, spending their lives permanently embedded in the mud. Lugworms are also typical inhabitants of muddy shores, sometimes present at densities exceeding 30,000 per hectare in very rich sediments. Like the worms of sandy shores, they construct U-shaped burrows through which they must pump a continuous stream of water to supply their oxygen requirements. These forty-centimetre-long annelids feed very much like terrestrial earthworms, ingesting quantities of

The enormous golden silk spider (**3**) of the Everglades spins a large web in order to ambush large airborne insects such as dragonflies. Out in the Atlantic Ocean the Portuguese man o'war (**5**) is equally predatory, the venom carried in the stinging cells of its tentacles being similar chemically to that of the cobra. Each man o'war is a colonial being made up of four different types of polyps – each one specialised either in floating, catching food, digestion or reproduction – and is thought to represent an evolutionary step towards the development of specific organs in higher animals.

mud, from which they extract organic material before excreting it and pushing it up to the surface to form the characteristic wormcasts which adorn European muddy shores.

The American equivalent of the lugworm on the Pacific coast is known as the innkeeper worm, on account of the numerous intertidal creatures which seek refuge in its spacious burrow when the shore is exposed. Instead of munching its way through the mud, it produces a mucus net, much in the same way as the parchment worm of sandy shores. The American

The iridescent purple gallinule (**6,7**) of the American tropics resembles the African and Asian jacanas in its lily-trotting habits.

ghost shrimp also extends the hospitality of its branched burrow to other shore-dwelling creatures, including pea crabs and gobies.

In many instances, muddy shores lie within the shelter of a river estuary, so creatures wishing to take advantage of the abundant supply of organic matter must also be able to tolerate regular changes in salinity as the proportion of fresh to salt water fluctuates with the tides. Strong currents are also common and the fine sediments brought down by the river and swirled from the bottom by the action of tides and currents

of these are the birds. In Britain, for example, it has been estimated that over one million wildfowl and two million waders utilise the estuarine habitats every year. Ducks and geese generally use the mudflats for roosting only, feeding on the eel-grass beds below the low water mark or in nearby fields, but waders both feed and roost here. The colloquial term 'waders' is used to describe long-legged, usually long-billed birds, many of which belong to the order Charadriiformes, which contains a total of sixteen families worldwide. Some waders frequent these mudflats only to replenish their fat

The white-bellied sea eagle (**4**) of Southeast Asia and Australasia hunts either by gliding low over the water or by keeping watch from an overhanging perch before swooping down to catch a fish or sea snake in its powerful hooked talons. Less agile than ospreys, these large, heavy raptors also take reptiles, mammals, birds and even carrion.

1

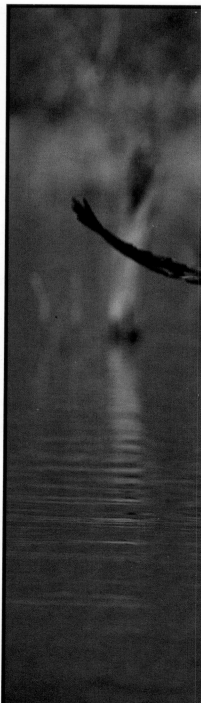

create high levels of turbidity. Estuaries are thus stressful habitats, unsuitable both for true freshwater and marine organisms; most creatures that can cope with these varying conditions are unable to live anywhere else.

Nevertheless, a zonation of life forms can be seen, largely dependent on the salt-tolerance of the species. Starfish and urchins, for example, can only advance into estuarine waters if their salinity is between thirty and thirty-five per cent, but some worms of marine origin, crabs and flatfish may be found where salinity is less than five per cent. Thus the upper levels of estuarine mudflats may support dense communities of red sludge worms, up to 250,000 per square metre, which bury their heads in the sediments to feed, and wave their tails in the water to create a current from which they can extract oxygen. A little lower down the shore, tiny spire shells munch their way through the surface sediments when the shore is covered with water, but are exposed as the ebbing tide drags the fine particles away. The lowest levels of the shore are populated by cockles and lugworms.

A large proportion of commercial fish species, especially flatfish such as plaice and dab, spend all or part of their lives in shallow estuarine waters, and many others while away their larval years there. Other fish have the ability to pass from salt water to fresh with no apparent modifications designed to cope with changing salinity. These include shad, lampreys, sturgeons, sea trout and salmon, which spend their adult lives in the sea but spawn in rivers, and eels, which adopt the reverse of this pattern.

Although some of these fish species are predatory, the main carnivores of the estuarine food chain originate from the terrestrial habitat, and by far the most abundant

2

3

4

reserves during the long migration between wintering and breeding grounds, but for others the plentiful food supply makes an estuary a good place to spend the winter.

Within any one community of waders there is a remarkable degree of specialisation in feeding techniques and requirements, so competition for food between species is minimal. In addition, differences in the times of day at which various species prefer to feed reduces competition for space. In this way, one area of mudflats can support literally thousands of birds. Most waders possess ear openings positioned much further forward than those in other birds, the better to detect vibrations in the mud, often the only sign that organisms are present below the surface.

Small, short-billed waders, such as the ringed plover, exploit surface-dwelling organisms of the tidal mudflats, especially the spire shells that can occur in densities of up to 42,000 per square metre. Those with long beaks are able to probe deeper into the mud – curlews and godwits, for example, are experts in hooking the fat-bodied lugworms from their burrows. Those with long legs are able to wade into the water to take advantage of any intertidal creatures that have not yet retreated into their burrows. Long-legged stilts and

5

The brilliant plumage of the scarlet ibis (**2,3**) is derived from a pigment obtained from its crustacean-based diet. These gregarious birds breed in mangrove swamps, forming colonies of thousands, and feed on open mudflats from Colombia to Brazil. The tip of the spoonbill's beak (**1**) is lined with sensitive tissue that detects small invertebrates in the shallow, murky water, prompting the bill to snap shut. Brown pelicans (**5**) often feed by diving rather than scooping gallons of water and fish into their extensible beaks from the surface.

avocets also feed in the sea, scything their flattened upcurved beaks from side to side through the surface sediments to disturb invertebrates; they also have slightly webbed feet so that if a wave catches them off balance, rather than floundering they are able to right themselves by swimming.

Fish-feeding birds, such as mergansers, cormorants, grebes, divers and herons, are also commonly seen in estuaries, along with otters, which frequently descend to the coast when their riverine habitats are frozen over during the winter. Seals are common denizens of estuaries all over the world, and other sea mammals which utilise such habitats are the herbivorous sea cows, or Sir*enia*, which inspired the ancient mermaid legends told by seafaring folk. Only four species occur in the world today: three species of manatee and the dugong of the coastal regions of the Indian and South Pacific Oceans. All of these are rare and declining in number. A fifth species, Steller's sea cow, was discovered in 1742 but was hunted to extinction by 1769. One of the manatees is a freshwater occupant of the rivers of the Orinoco and Amazon basins, but the others – the West Indian and West African species – are found mainly in the coastal waters of their respective continents.

The influence of the sea may extend far upriver, the tides bringing an influx of sea water twice daily, so that the upper regions of the estuary are still part of the coastal ecosystem. Here, where sediments brought down by the river and washed ashore by the waves

grass. Cord grasses are dominant over large stretches of intertidal saltmarsh throughout the northern hemisphere. They fulfil a stabilising function similar to that of marram grass on sand dunes, and, like the undersea meadows, support a distinct, complex community. Black snails graze on the algae which coat the stems and leaves, and purple marsh crabs feed on the leaves themselves. Marsh snails ascend the tallest stalks to avoid the incoming tide as they are equipped with the lungs of terrestrial molluscs and cannot remain submerged for more than an hour.

When the level of the saltmarsh has risen to such an extent that it is only sporadically submerged, sea purslane, sea plantain, sea club-rush and saltmarsh rushes make an appearance, along with colourful sea

1

Sealions (**2**) live in enormous mixed herds for most of the year, dispersing at the onset of the breeding season, when the bulls select territories and harems, defending them diligently against intruders. California sealions (**8**) occupy the coasts of California, Japan and the Galápagos Islands, and South American sealions (**6**) breed right around the southern two-thirds of that continent. Harbour, or common, seals (**3,5**) utilise the Pacific and Atlantic coasts, and the larger grey seals, with their pure white pups (**7**), are particularly abundant on the northern island groups of Great Britain.

2

have been consolidating over a longer period of time, the level of the land is slightly higher and tidal incursions are very shallow. Halophytic plants may start to colonise these mudflats, and so a saltmarsh is born.

One of the most common pioneer species of European mudflats is marsh samphire, or glasswort, which has fleshy stems and leaves that are much reduced in order to conserve its tissue fluids in the saline environment. Around its upright stems more sediment accumulates, enabling other species to become established. Annual seablite is one of the first plants to join the glasswort beds, together with halophytic grasses, such as cord grass and sea manna

asters, sea thrift and sea lavender. Eventually a saltmarsh will rise to such a height that the sea no longer reaches it. At this stage, the dominance of the halophytic species is reduced and brackish grasslands develop, often used as pasture for cattle.

Throughout the temperate regions of the world, saltmarshes and mudflats are some of the richest habitats in terms of both species diversity and numbers of individuals within each species, especially with respect to the bird life they support; for this reason they are among the most precious parts of our natural heritage. In Europe alone such complexes of salt and freshwater wetlands at the mouths of great rivers

Killer whales (**1**) are almost global in their distribution and are small enough to frequent coastal waters; they have even been known to snatch seals from the beach.

Harvest mice (**4**) are usually thought of as residents of grasslands, both natural and planted, but they also frequently utilise coastal saltmarsh habitats. The smallest of all European rodents, the harvest mouse is distributed from the British Isles to Japan.

4

3

5

6

7

attract tens of thousands of birds every year, some regularly to feed, others just passing through, but the majority to nest and rear their young on the rich harvest of the coastal marshes. The Camargue, at the junction of the Rhône and the Mediterranean, supports some 10,000 flamingos, 800 pairs of avocets and 400 pairs of purple herons among the 200 or so species of birds that choose to breed and feed here.

The Doñana marshlands at the mouth of the great Guadalquivir on the south Atlantic coast of Spain regularly play host to some 125 species of breeding bird, including spoonbills, cattle egrets, herons and storks, as well as the rare marbled teal and white-

8

headed duck. The Danube delta on the Black Sea encompasses some 5,000 square kilometres of fresh, brackish and saltwater habitats, including the largest expanse of reedbeds in the world. It is home to about a hundred fish species, including the sturgeon, and more than 150 species of bird. The white pelican and glossy ibis, for example, breed in the Danube delta and in only a handful of other sites in Europe. The delta is also home to more common species, including night herons, little egrets, penduline tits, black-necked grebes and pygmy cormorants.

The mangrove swamp is the tropical equivalent of a temperate saltmarsh. Although the resemblance in terms of vegetation is slight, mangroves, too, are intertidal species, able to withstand variable salinity. They are found fringing more than half of all tropical shores, extending into subtropical regions as far south as New Zealand and northwards to Japan and the Red Sea. The coastline of Africa supports about three and a half million hectares of mangrove swamps, and that of tropical Asia more than six million hectares, with a similar extent around the shores of South America. The swamps are dominated by trees, the actual species of tree depending upon its proximity to the sea.

Where silt accumulates from river-borne sediments on sheltered shores, the land is able to advance seawards due to the progressive colonisation of the

1

2

There are over 630 species of true tree frog in the world, distinguished by the presence of disc-like suckers at the tip of each digit. Many species, such as the green tree frog *Hyla cinerea* (**3**) are found in dense coastal wetland vegetation.

3

A breeding pair of white-bellied sea eagles (**1**,**4**) will share the task of nest-building, usually constructing their abode on a rocky ledge or in a tree. The female lays two to three eggs, from which, in most years, only one eaglet will be successfully reared, though when food supplies are plentiful all the nestlings may reach maturity. The juvenile birds leave the nest after two months, but continue to return to their parents to supplement their diet for a further six months. Brown pelicans (**2**) were once common along the Pacific, Gulf and Southeast Atlantic coasts of the United States, but their numbers have decreased in recent years due to the indiscriminate use of pesticides. The magnificent American egret (**7**) still breeds regularly in the lush swamplands of Florida's Everglades region.

mudflats by mangrove seedlings. The pioneer species, which take hold just above the level of the lowest tides, are *Sonneratia* and *Avicennia* , commonly known as black mangrove trees. As the trees grow they develop an array of slender aerial roots which protrude from the mud. Tiny pores on the surface of these roots absorb atmospheric oxygen to supply the subterranean roots, which would otherwise suffocate in the closely packed sediments. These roots also serve to trap more silt brought in with each successive tide.

Gradually the level of the swamp rises, inundations of sea water become less frequent and the pioneer mangroves are replaced by other species. *Rhizophora* the red mangrove, supported by its buttress-like prop roots, and *Brugiera*, typified by long, looping roots which undulate above and below the sediments, dominate the middle sections of the swamp. Where the influence of the sea is minimal, at the back of the swamp, the mangroves give way to species such as

nipa palms, figs, buttonwood, mahogany, bromeliads, ferns and other such typical components of coastal tropical forests.

There are about sixty species of tree which are known collectively as mangroves. They belong to several different families but have taken on a similar appearance on account of the habitat in which they live. Mangroves cope admirably with the stresses of an intertidal lifestyle, avoiding the accumulation of excess salts in their tissues in a number of ways. Some have special salt-excreting glands on the leaves, others deposit salts in a few old leaves which are then shed along with their toxic cargo, and others evade the problem at an earlier stage by possessing a membrane around the roots, which allows water to pass through but filters out salt molecules. For the most part they are evergreen species, their leaves having thick cuticles with the stomata buried in pits in order to retain water within their tissues.

process of making a great evolutionary leap from sea to land, following the same route as the first amphibians.

Mudskippers have stumpy pectoral fins that enable them to 'walk' across the exposed mud, and some even have pelvic fins fused together to form suckers so that they can cling to vertical prop roots or tree trunks. Unlike most fish, they are able to breathe whilst out of the water by carrying a mixture of air and water in their gill chambers. Unfortunately, if they open their mouths to feed, this fluid comes gushing out and the mudskipper must dash back to the edge of the sea to gulp another mouthful. Their eyes are located on the tops of their heads so that, even if taking a rest in shallow water with their bodies fully submerged, they are able to keep an eye on what is happening on the nearby shore.

Like the mudskippers, many of the mangrove crabs are able to exist equally well on land or in the sea.

Electric rays (**8**) are bottom-dwelling cartilaginous fish found in shallow tropical and sub-tropical waters. They have flat, almost circular bodies which bear electrical organs on either side capable of producing a shock of up to 300 volts, sufficient to stun a good-sized fish.

Many species of mangrove produce seeds which germinate whilst still attached to the tree. Only when the seeds have produced roots up to sixty centimetres long, and thus have a reasonable chance of survival, do they fall from the tree. If the tide is low, the root will penetrate the mud like a spear, anchoring the young mangrove firmly in place; if the tide is high, the seedlings fall into the water and float horizontally on the surface of the sea. In this way they may be carried for many kilometres before encountering bare mud in which they can flourish.

By virtue of their intermediate position between sea and land, mangrove swamps are populated by a wide variety of both terrestrial and marine organisms, many of which are amphibious and thus able to survive in both habitats. The most noticeable amphibious creatures of the mangrove swamp are the mudskippers, so named for their ability to hop across the mud using their tails for propulsion. They appear to be in the

Red mangroves (**5**), with their arching prop-roots, are found throughout the tidal zones of the tropics. The mangrove swamps of Southeast Asia are the haunt of the fishing cat (**6**), which reaches over a metre in length. These felines are usually seen hooking fish out of the murky waters with their paws, and it is not known whether they ever pursue their prey by swimming or diving.

Ghost crabs, marsh crabs and hermit crabs are abundant, but the most typical is the fiddler crab. The male of this species has a lopsided appearance due to the development of one claw to a size out of all proportion to its body. It has been suggested that the function of such an appendage is not so much for defence as to attract a mate. Each male has a different means of gesturing to the female of his choice, some beckon, some wave and others embark on a complicated sequence of semaphore signals, but the end result is usually the same and the female accompanies him to the privacy of his burrow.

Mangrove reptiles range from loggerhead turtles to diamond-back terrapins; from arboreal snakes, such as

1

4

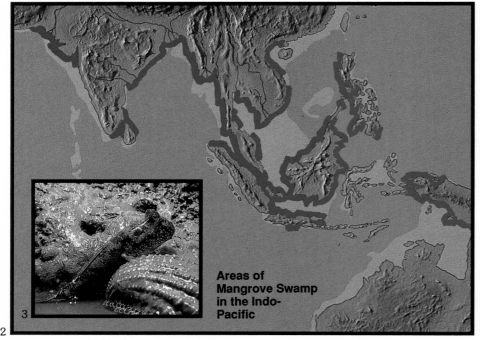

Areas of Mangrove Swamp in the Indo-Pacific

3

2

Wagler's pit viper and the mangrove snake, to the vividly striped marine coral snakes and water moccasins, and from the cannibalistic alligator of Florida's Everglades to the estuarine crocodile of southeast Asia and Australia. The latter is the largest of its kind, sometimes reaching a length of six metres or more. But the birds of the mangrove swamps attract most of the attention, partly because of their large size, but also because of their brilliantly coloured plumage. During the course of the year in the coastal regions of the Everglades, the mangroves are home to such splendid long-legged birds as roseate spoonbills, great blue herons, glossy ibis, snowy egrets, green-backed herons and the now rare wood stork, as well as brown pelicans, ospreys and southern bald eagles, most of which are fish-eating species.

On Borneo, the Baram and Limbang flood plain grows by some thirty metres per year through progressive sediment accumulation and mangrove development, leaving coastal towns far behind in the process. The mangrove swamps of the island are some of the most diverse in the world, containing about thirty different species of mangrove tree. They are the only place on this planet where the bizarre proboscis monkeys live, feeding only on *Sonneratia* leaves. Other monkey species that are typical of the world's mangrove swamps include the silvered and dusky langurs of southeast Asia and the long-tailed macaque, which may wade up to its waist in water searching for the crabs on which it feeds. But on the whole, mammals are few and far between in the mangroves, with the exception of otters, crab-eating raccoons, small arboreal cats and marine mammals, such as the West Indian manatee.

Of all the coastal ecosystems of the world the coral reef represents the outermost stage in the reclamation

5

6

7

Florida cooters (**10**) are among eighty-five species of omnivorous pond and river turtles, sharing their habitat with American white pelicans (**9**) in winter, although these huge birds breed inland. Swamp cypresses (**4**) tower over the maze of waterways that dissects these swamplands, and anhingas (**11**), also known as snake-birds or darters, perch above the water, their long necks kinked and wings outstretched.

of land from the sea. In addition, it is undoubtedly the richest of all marine habitats in terms of species diversity, often compared to the tropical rainforests, which are the most diverse terrestrial ecosystems. The largest coral reef in the world is the Great Barrier Reef. It stretches for about 2,100 kilometres along the northeastern coast of Australia, extending on average for 300 kilometres outwards from the mainland into the ocean, and has a total area of some 2 160,000 square kilometres. It is estimated to contain over 3,000 different animal species, many in huge numbers, of which about 350 are reef-forming corals. In addition, the isolated cays, or built-up atolls, support the largest seabird colonies in the world.

trap the zooplankton on which it feeds. Once it has reached a certain size the polyp will produce one or more buds, which in turn will develop into new polyps. In the majority of cases the new polyp develops on top of the original and kills it, so that a large mass of coral may bear only a thin skin of living creatures, supported by the empty chambers of their ancestors. In this way the characteristic stag's-horn, brain, finger and star corals achieve the distinctive shapes by which they are identified.

Corals can grow at a rate of up to eight centimetres per year in favourable conditions. The majority of them have a symbiotic relationship with tiny green algae known as zooxanthellae. It is still not certain exactly

Mudskippers (**3**) are among the best-known inhabitants of the mangrove swamps of the Indo-Pacific region, whose distribution is markedly coastal (**2**). Other inhabitants of these tidal thickets are the arboreal golden-banded mangrove snake (**7**), a wealth of semi-terrestrial crabs belonging to the Grapsidae (**8**) and the silvered langur (**1**), a vegetarian leaf monkey. Indo-Malayan monitors (**5**) partly fill the niche of the predatory mammals in the absence of such creatures in the Australasian region, where pied stilts (**6**) have the same black and white plumage and long legs as their northern hemisphere relatives.

8

9

10

Calcareous coral reefs thrive only in warm, unpolluted waters, ranging from about 18°C to 35°C in temperature, and are thus limited to a zone which lies roughly between the latitudes of thirty degrees both north and south of the equator. They are rare, however, along the western coasts of the continents, where the continental shelf falls away rapidly and cold currents frequently well up from the deeps. There are three main types of reef: fringing reefs, which extend directly from the shore; barrier reefs, which run parallel to the shore but are separated from it by a stretch of water; and atolls, which are circular reefs enclosing a central lagoon, usually perched on the summits of submerged volcanic islands.

Coral reefs are composed almost entirely of the calcareous exoskeletons of the coral polyps, each of which resembles a tiny sea anemone. When the larval forms of these creatures encounter a suitable site they metamorphose into tiny hydroids, each of which immediately secretes a protective limestone cup around itself, from which it extends its tentacles into the sea to

11

what function these algae serve, but they are thought to provide the coral polyp with certain vitamins and hormones which it cannot manufacture for itself. They may also facilitate the secretion of its calcareous exoskeleton and absorb waste products. There may be as many as 30,000 zooxanthellae in a single cubic millimetre of living animal tissue, weighing up to three times as much as the polyp itself.

Because green algae manufacture energy by means of photosynthesis, a process which requires sunlight, reefs cannot develop below a maximum depth of about 150 metres. Turbidity in the waters also reduces the amount of sunlight which filters through, so that reefs cannot survive where large quantities of silt are washed into the sea by rivers. For this reason they are often associated with mangrove swamps, which act as efficient filter systems for these sediments.

Apart from the corals, other plankton-feeding organisms make use of the constant influx of fresh food in the waves and currents that wash the reef. Some sea anemones are able to grow over a metre in diameter on this rich diet, and the giant clam of the Indian and Pacific Ocean reefs can weigh up to 250 kilogrammes. The seaslugs of the reef, which feed mostly on anemones and sponges, are a far cry from their dull-coloured terrestrial cousins, sporting all manner of colours and patterns.

There are a number of fish species which feed directly on the corals themselves, including such specialists as the leather-jacket fish, which 'sucks' the

polyps from their chambers. The parrot fish is not so fastidious in its eating habits, nibbling away chunks of coral, limestone and all with its horny, beak-like mouth. Its powerful throat 'teeth' grind the coral into fine particles by muscular contractions and the edible parts are absorbed by the stomach. The undigested calcareous fragments are returned to the reef via the anus, accounting to some extent for the fine coral sands which surround the sedentary reef organisms. The various species of butterfly fish, impossibly flattened from side to side so that they are almost invisible from

Tropical coral-reefs (**10**) are home to a dazzling array of fish species, including the symbiotic clownfish *Amphipryon percula* (**5**) and *A. xanthurus* (**6**), the leopard scat (**1**), spotted to resemble coral polyps, and the turkey fish (**3**) with its greatly elongated pectoral fin-rays, all of which are found in Indonesian waters. More cosmopolitan species include the squirrel fish (**2**), with a worldwide tropical distribution, while the Californian coast supports such species as the well-camouflaged kelpfish (**4**), the spectacular blue-banded goby (**7**) and the bizarre, crevice-dwelling blue ribbon-eel (**11**). Sea fans (**8**) are made up of eight-tentacled polyps which, unlike reef-building corals, produce an internal skeleton.

front and rear, have long, pointed snouts with which to probe the reef crevices for invertebrates sheltering within.

Coral reefs support about one third of all known fish species in the world. The vast majority are brightly coloured and thus never venture far from the reef as they are much too conspicuous in the open ocean. These splendid liveries are not purely for display. In many instances they help the fish to recognise other individuals of the same species for reproductive purposes, or to warn off rivals from their breeding territory. In other cases red, yellow and black are employed to warn potential predators that these fish are poisonous, although other species mimic these patterns and are able to avoid being eaten despite the fact that they are not toxic. The cleaner wrasse is an example of a fish which feeds on the parasitic organisms of larger fish, a habit which benefits both parties. Its colours act as a sort of safe-conduct pass.

9

10

11

As well as being among the most diverse habitats of the world, coral reefs are also some of the most fragile. The deforestation of a tropical mountain island will often result in erosion of large quantities of topsoil. These sediments find their way into rivers and eventually to the sea, where they will quickly smother the coastal reef beneath a layer of fine silt, a process which can also occur following the dredging of a harbour. Reef limestone is mined in some parts of the world for use in the construction industry, which causes untold damage to the whole community. Increasing tourism

on tropical islands is provoking a demand for souvenirs, and many attractive corals and molluscs are being overharvested to supply the kiosks. Even glass-bottomed boats, used to catch a glimpse of the wonder of reef life, can cause problems in terms of pollution and wave damage.

The sudden population explosion of the crown-of-thorns starfish in the Pacific reefs in the 1960s has also been attributed to man's upsetting the equilibrium of the reef in some way, although no-one is quite sure how. This enormous, sixteen-armed starfish literally eats its way through the calcareous structure of the reef, feeding on the coral polyps but opening up large holes which are then susceptible to wave erosion. By 1969 it had laid bare more than 360 kilometres of Australian reefs.

But it is not only the coral reefs that are being damaged. All over the world coastlines are suffering at the hands of man. Mangroves are cut down for timber and decimated by peat extraction; estuaries are flooded with nitrates from fertilisers applied to the arable lands of the northern hemisphere; coastal power stations warm the shallow seas to the point where the once diverse communities of saltmarsh and rocky shore alike are reduced to a few hardy species, and the many tourist resorts of the world pump their sewage untreated into the sea. The very existence of such resorts indicates man's age-old fascination with the interface between land and sea, but its sparkling coastal seas and unspoilt beaches are in danger of disappearing forever.

12

The crown of thorns starfish (9,12) feeds by everting its stomach, wrapping it around a lump of coral and sucking out the living polyps. Periodic plagues of these enormous echinoderms have devastated vast expanses of reef, particularly around Australia.

· CHAPTER 9 ·
FRESH WATER

Although the surface fresh water of this planet covers
less than two per cent of its land area, it supports a
phenomenal richness and abundance of aquatic wildlife
and, indeed, is ultimately responsible for the existence of
all terrestrial life-forms as well. From the ancient depths
of Lake Baikal to the garden pond, from the mighty
reaches of Nile and Amazon to streams which are born
of mountain snows, from seasonal marshes to
permanently flooded river deltas and great expanses of
peat, the range of habitats provided by fresh water is
vast. Truly aquatic creatures range from microscopic
planktonic organisms to the bloodthirsty piranhas of the
Amazon, while a whole host of mammals, water birds
and amphibians take advantage of fresh water for
feeding or breeding.

Right: a sea of rose-pink flamingos, (top) open-bill storks
silhouetted against the sunset and (above) a
newly-emerged swallowtail.

Fresh water is so called because it is relatively pure compared to that of the oceans; even the hardest fresh water contains only a half of a per cent of dissolved mineral elements, whereas ocean salinity is approximately thirty-five per cent. The seas and oceans contain over ninety-seven per cent of all the world's water, and a further two per cent is currently frozen in the permanent icecaps and glaciers of the world. The remaining fraction, less than one per cent, represents the total of all liquid fresh water on earth, and even of this small amount about two-thirds lies underground at any one time.

And yet the majority of terrestrial life-forms, including human beings, are dependent upon a regular intake of fresh water. Some creatures, including about a quarter of all the world's fish species, never leave their freshwater habitats, while others, such as frogs and newts, spend much of their lives in the water, especially during their larval stages. Turtles and crocodilians feed in the water but rest and lay their eggs on the shore, whilst a few large mammals, such as hippopotamuses and capybaras, retreat to the water for relief from the sun and refuge from terrestrial predators.

1

Various species of swallowtail butterfly (**4**) are found across the temperate northern hemisphere, often associated with freshwater wetlands as the larvae thrive on marsh umbellifers. More local is the Mallorcan midwife toad (**1**), endemic to permanent pools in mountain streams on this island, where it is out of reach of the introduced viperine snake, its main enemy. Only a few thousand wild water buffalo (**6**) still exist in southeast Asia, where they live in herds of ten to twenty and favour swampy grassland.

2

One reason why such a small amount of water is able to support so many living organisms is the existence of the hydrological cycle. All large waterbodies, including the oceans, are subject to the evaporation of their surface waters during hot weather. The water molecules so removed enter the atmosphere as vapour, forming clouds, later to be returned to the earth as precipitation – that is, rain, snow or dew. For the most part water, which falls as precipitation percolates through the soil until it encounters a layer of impervious rock, and thus forms vast underground aquifers. Where this ground water, as it is called, comes into contact with the surface, springs occur, and these often feed streams.

Several streams may amalgamate to form a river, which, in accordance with the laws of gravity, will flow downhill until it reaches its lowest point. In many cases, this is the sea, and thus the cycle is complete. The important thing to note in this simplified sketch is that the oceans are an integral part of the hydrological cycle, making a huge volume of water available for circulation around the earth, even though the amount of fresh water present at any one time is extremely small.

Wetlands is the collective term for those habitats in which water is a predominant feature, although it also encompasses such saltwater ecosystems as estuaries

The Carolina wood duck (**2**) is a North American species that breeds in natural tree holes, or those vacated by woodpeckers, in deciduous or mixed woodlands, always close to shallow lakes or marshlands.

3

Freshwater plants range from Australia's magnificent smooth-barked river red gums (**3**) to the water hyacinth (**5**), a rampant aquatic weed. A native of South America, the latter has been widely introduced and reproduces very rapidly, blocking major waterways all over the world.

4

5

and mangrove swamps. Freshwater wetlands can be divided into two major types of ecosystem: those through which the water is constantly moving – lotic habitats, such as springs, streams and rivers – and those in which it is more or less still – lentic habitats, ranging from seasonal swamps and marshes to ponds, lakes and peat bogs. The distinction between a stream and a river is largely one of size, as is that between a pond and a lake, although ponds on the whole are shallower and may dry up for part of the year. Swamps and marshes are less easily defined, but generally speaking swamps are permanently flooded areas, whilst marshes are only seasonally wet. In many cases

6

these wetlands are points of transition between aquatic and terrestrial habitats, and as such support a rich assembly of plants and animals typical of each environment.

Apart from the obvious link of the hydrological cycle, these two types of freshwater biome are also connected in another way. Freshwater rivers receive a continuous supply of nutrients from the land, initially entering the system via ground water or surface runoff following precipitation. Mineral elements are added through chemical erosion by turbulent mountain streams and organic matter by the input of fallen leaves from waterside vegetation. Yet most of these resources are washed straight through the system without being

The white swans (2) are essentially northern hemisphere birds which require aquatic habitats at all times of year. Trumpeter (3,4) and whistling swans represent the New World species and mute (5) and whooper swans breed across much of Eurasia.

Whooper Swan

Mute Swan

Bewick's Swan

Trumpeter Swan

Whistling Swan

used, eventually serving to enrich oceans and lakes, which are thus the more productive habitats.

Broadly speaking, the result of this downward flow of nutrients and sediments is that all lakes are in the process of becoming dry land – although this may take many thousands of years in the case of the largest waterbodies. The water becomes shallower and vegetation is able to colonise the rising sediments, with the result that lakes gradually give way to swamps and then marshes. In many cases, the sediments eventually become dry enough for tree species to invade – alder and sallows in temperate climates, followed by the climax woodland of the region, usually ash and oak. Ultimately the lake may become little more than an expanse of wet woodland in a low-lying depression through which the original river still flows.

Wetland plants are commonly known as hydrophytic species – those which favour watery conditions and are

Some ornithologists consider Bewick's swan (6), which breeds on the northernmost edges of the tundra, to be a Eurasian subspecies of the whistling swan.

7

8

Crested coots (**7**) are essentially birds of African lake margins although there is also a small resident population in the Doñana marshes in southwest Spain. The closely related common coot (**9**) feeds mainly on aquatic plants, which it obtains by diving to depths of up to seven metres; the feathers are flattened to exclude air before the coot leaps upwards to gain propulsion for its dive, assisted underwater by its strong, webbed feet. Marsh marigolds (**8**) were once common plants of wet grasslands and woodlands across Europe, but have now been lost in many areas following drainage for agriculture.

adapted to cope with the problems imposed by such aquatic and semi-aquatic environments. All plants require carbon dioxide in order to photosynthesise and oxygen for respiration, but the concentration of oxygen in water is at most only three per cent of that present in the atmosphere, although carbon dioxide is more soluble. Completely submerged plants are forced to satisfy their gaseous requirements by absorbing these gases from the surrounding water. Since oxygen availability is so low, however, some species have evolved a method whereby they retain the oxygen produced during the process of photosynthesis in hollow chambers known as lacunae within their stems, thus supplying the respiratory needs of the plant from within.

Waterlogged soils are also a feature of many wetlands. The lack of airspaces in the closely-packed sediments creates anaerobic conditions, depriving the roots of a ready supply of oxygen. Where soils are waterlogged for only part of the year, especially in high latitudes during the winter, some hydrophytic plants opt to avoid such stressful conditions by becoming dormant. Other species produce lateral roots close to the better-oxygenated soil surface, and some, such as the bald or swamp cypress, may produce aerial roots – pneumatophores – which obtain their oxygen from the atmosphere rather than the soil.

9

Herbaceous plants, however, adopt a different approach to supplying their roots with oxygen. Many have roots and stems filled with aerenchyma – thin-walled cells with large air spaces so that oxygen entering the plant through the stomata of aerial leaves and stems can be transported downwards to supply the respiratory needs of the roots growing in anoxic soils. Other wetland plants have evolved a method whereby their roots can respire anaerobically, that is, without oxygen, for short periods. Among such plants are the waterside iris known as the yellow flag, some waterlilies and reedmace. Light is also a limiting factor with regard to the growth of hydrophytes, as they can only photosynthesise at depths which receive at least three per cent of full sunlight. This may be less than a metre in silt-laden waters, though the clearest lakes support vegetation to a depth of over ten metres.

The distribution of the waterbuck across Africa is, predictably, governed by the presence of water, as these bovids never stray far from wetland areas and retreat into the water when danger threatens. The Defassa waterbuck (**2**) is distinguished from the common waterbuck by its lack of a crescent-shaped mark on the rump.

to maintain their position at the surface, the leaves are large and flat, often circular or oval in shape, with well-developed aerenchyma for buoyancy. Rough waters would quickly swamp such leaves, so typical floating plants – waterlilies and lotus flowers, pondweeds and frogbit – grow only in still backwaters and sheltered ponds. They vary considerably in shape and size: the leaves of the giant or royal waterlily of the Amazon may be two metres in diameter, those of the tiny duckweeds are often less than five millimetres across, whilst the related wolffia is probably the smallest flowering plant in the world, consisting of a globular thallus – no roots or leaves – which may be less than a millimetre across. The upper surfaces of such floating leaves possess a waxy cuticle to displace rain water, and also bear stomata, essential for oxygen intake from the atmosphere.

Other floating plants, such as the water-hyacinth, have more upright leaves bearing inflated, bladderlike petioles to keep them at the surface. The tropical

In their natural habitat in North America, Canada geese (**1**) are highly migratory birds, but where they have been introduced into Europe they seem quite content to lead a settled existence, some even spending their whole lives on city lakes, providing sufficient food is available. They mate for life and both parents care for the five or six fluffy yellow chicks.

3

4

5

Submerged plants are normally attached to the river or lake bottom, but, although the vegetative parts are completely under water, many species, such as the carnivorous bladderworts and water lobelia, produce flowering stems that rise above the surface, so they may be pollinated by wind or insects. The hornworts, however, are perhaps more truly aquatic than any other freshwater vascular plant. They are not rooted in the bottom sediments, but float freely beneath the surface and flower under water. The mature anthers rise to the surface attached to a small float, where they dehisce in the drying air and release their pollen. The pollen grains then drift down through the water to fertilise the elongated stigmas of the female flowers.

Floating plants are those whose leaves occupy the interface between air and water, and as such have to cope with two sets of environmental stresses. In order

The abundance of life-giving water in marshlands engenders a luxuriance of vegetation not seen in other habitats. Globeflowers (**5**), among the more attractive members of the buttercup family, and purple-loosestrife (**4**) both thrive in wet meadows and alongside natural watercourses throughout much of Europe, while the insectivorous sundew (**3**) is essentially a species found in more acidic peatlands.

The diversity of animal life in the world's wetlands is phenomenal, ranging from the tiny, five-centimetre-long European tree frog (**7**), well equipped for climbing with its long limbs and adhesive pads on fingers and toes, to the four-tonne great Indian rhinoceros (**9**), which stands 210 centimetres at the shoulder and sports a single horn up to sixty centimetres long. The American egret (**1**), also known as the great white heron, is an elegant bird of the Florida Everglades, stalking slowly through the water and jabbing with its lethal beak at small fish and amphibians.

water-cabbage consists of a floating rosette of leaves, each of which has a spongy swelling on its underside which acts as a float, while its upper surfaces are densely clothed in fine hairs to repel water. Some of these species, such as pondweeds and waterlilies, are rooted in the bottom sediments and must therefore have long, flexible stalks to allow for changing water levels, whilst others are free-floating at the surface, such as frogbit and water soldier.

Many floating aquatic plants reproduce both through bearing large, attractive flowers and also vegetatively in various ways – by means of stolons, such as the water cabbage; by the break up of the parent into individual young plants, as in the water fern; by the production of axillary buds in times of plenty, each of which is capable of developing into a new plant when shed, as in the Canadian pondweed; or by the budding of new plants from the parent thallus, as is the case in the duckweeds. In suitable conditions these plants can spread very rapidly, choking watercourses with a luxuriant carpet of vegetation which not only restricts navigation, but also prevents light from reaching the lower layers of the pond or river and uses up all the available oxygen, so that other plant and animal life is seriously depleted.

Marsh marigolds, or kingcups (**6**), are found throughout Arctic and northern temperate wetlands. Comfrey (**4**), also known as bone-set, was a common component of ancient herbal remedies.

Unlike the algal seaweeds which dominate the seashore, most freshwater plants possess vestigial stomata and cuticles, indications that they were once land plants which have returned to the aquatic habitat. As the surrounding water provides support for aquatic plants, woody tissues are unnecessary, while the environment also cancels the need for the complex water-transporting tissues required by land plants.

The third type of aquatic plant is that which grows in shallow waters, having roots beneath the surface and foliage in the air, often seen in dense stands in swamps and on the margins of lakes and rivers. These species, such as yellow flag, rushes, papyrus, reeds and bullrushes, differ little from land plants in that they possess well-developed vascular tissues and

6

7

8

9

10

Floating aquatic plants include the Amazon water-lily (**2**) of tropical South American rivers, with its two-metre-diameter leaves, and the diminutive yellow Rocky Mountain pondlily (**3**). The water horsetail (**8**) is one of sixteen species of these primitive plants, found all over the world except in Australasia.

strengthening fibres since they lack the support that water affords submerged plants. Aerenchyma for gaseous exchange is still present, however, in young plants that spend the first stage of their lives immersed in shallow waters and in those that root in anoxic sediments.

Exceeding 6,400 kilometres in length and draining an area of more than seven million square kilometres, the Amazon is undoubtedly the greatest river on earth. Although a few hundred kilometres shorter than the Nile, the volume of water which flows between the banks of its many tributaries is unequalled by any other

watercourse. From the trickling headwaters which rise in the snow-covered peaks of the Peruvian Andes, less than 200 kilometres from the Pacific coast, the Amazon carves a path through the tropical rainforests, gathering strength from some 1,100 tributaries along the way. By the time it reaches the Atlantic Ocean, the mouth of this great river is over 3,000 kilometres broad and the force of its outpouring such that fresh waters predominate over saline as far as 180 kilometres from the coast. The lower reaches of the river, if viewed from the air, show a classic tangled maze of waterways known as 'braiding'.

Papyrus, or paper reed (**5,10**), is native to tropical Africa but has been widely introduced elsewhere. Its stems, which may be up to five metres long, were used in ancient times to make paper.

This short profile of the Amazon serves to identify some of the essential characteristics of all the world's rivers, whether only a few hundred metres or thousands of kilometres in length. The upper reaches of a river are turbulent and clear, rich in oxygen but poor in nutrients. The narrow bed and fast-flowing waters which typify such 'young' rivers possess considerable erosive abilities, both by chemical weathering of the rocks they pass over and by the abrasive action of rock particles – and even small boulders when the river is especially full – against the sides and bottom of the channel. Such erosive rivers are capable of slicing through the rock over the course of the years, resulting in the formation of gorges and defiles; one dramatic example of this is found along the Colorado River as it snakes through the world's largest gorge, the Grand Canyon, which is almost two kilometres deep in places. Where the water comes into contact with a band of resistant rock a waterfall may form, one of the most impressive examples being Niagara Falls in North America, although Venezuela's Angel Falls have the distinction of being the highest in the world, as their waters plummet for over a kilometre.

1

The shy, skulking bittern (1) of Europe and the aptly-named African goliath heron (6), standing up to 120 centimetres tall, are typical birds of the wetland fringe, whereas grebes, such as the red-necked species (2) of eastern Europe and western Asia, are better adapted to the open water. The coypu (3), a South American rodent, is a highly specialised aquatic mammal with webbed hind feet and is able to remain underwater for up to five minutes.

2

3

4

5

African elephants (4) utilise flooded areas to feed and protect themselves against sun and insects. Blue damselflies (7) are distinguished from true dragonflies by their habit of folding their wings vertically at rest. Marsh horsetails (5) reproduce by means of sporangia borne terminally on stems lacking in chlorophyll.

As the young river travels down towards the sea it gathers water from subsidiary streams. The nutrients which it has gleaned from the bedrock in passing, together with those derived from the dead leaves of riverside vegetation, enrich the waters and allow more life to flourish. At the same time the energy involved in carrying this increasing sediment load reduces the potential erosive capacity of the waters. The river at this stage is known as 'mature', yet for all its tranquil appearance it actually flows more swiftly than the turbulent headwaters, largely because its greater volume and wider channel engender less friction between water and land. Where the mature river meets an obstruction it will curve elegantly away, creating a wide loop as it resumes its original seaward course. In this way a series of meanders is formed. A renewed surge of energy on the part of the river – often provoked by heavy rain in the mountains – may create an isolated oxbow lake as it cuts a new course across the neck of the meander.

The lowest reaches of the river, approaching the sea, often cross land with very little gradient. The old river loses strength and its slow-flowing waters can no

6

7

8

longer carry their substantial sediment load, which is deposited in the most lethargic parts of the river. In a suitable low-lying depression, the riverine labyrinth dissecting the floodplain can comprise thousands of square kilometres, as in the case of the Okovango swamp in Botswana, the inland culmination of the Cubango River, which rises in the hills of Angola's Bié plateau, and never reaches the sea.

It is thus obvious that a single river, from its turbulent headwaters to the placid floodplain, provides a whole range of habitats to be colonised by creatures with vastly different ecological requirements. The main limiting factor to life in a river is the strength of its current, which is itself dependent upon gradient and

9

Nile crocodiles (**9**) lay up to ninety eggs in the dry season so that their hatching period coincides with the rains four months later. The Nile monitor lizard (**8**) lays its eggs within a rain-softened termite mound, ensuring food and shelter for the young when they hatch.

the amount of precipitation. Temperature is also a consideration. Although in itself this is unlikely to rise or drop to the level where it restricts life, warm waters carry less diffused oxygen than cool ones. Where factories or power stations utilise river water as a coolant and release it back into the watercourse after use, severe oxygen depletion can occur, killing whole communities of riverine organisms.

A fast-flowing mountain stream is usually devoid of vascular plants, which are unable to resist the strong current and constant grinding of pebbles; instead, lower plants form the basis of the food chain. Unicellular organisms such as diatoms and blue-green algae form a thin coating on the pebbles of the stream bed, whilst some mosses, ferns and liverworts that are able to grow submerged attach themselves with their customary tenacity to large, stationary rocks near the surface where sufficient light is available for photosynthesis.

Some fish, such as trout in the highest streams and grayling and mullet lower down, swim continuously upstream at a rate equivalent to the downwards flow of the water, thus maintaining their position in the same stretch of river. The top speed of such fish must always exceed that of the current so that, should danger threaten, they are able to increase their efforts briefly

1

2

3

Pelicans are among the largest of birds, with strong, webbed feet and long necks. Their most characteristic feature, however, are their enormous beaks, the lower mandible carrying a nine-litre pouch which acts as a fishing net in the majority of species, such as the Asian spotted-billed pelican (**8**). The brown pelican (**1**), on the other hand, collects its food not from the surface of the water but by diving from a great height. Black-headed gulls (**4,5**), although considered to be coastal birds, often breed and winter inland on freshwater lakes, bogs and marshes. The crepuscular Australian black duck (**3**), an inhabitant of densely vegetated, permanent lakes, is the continent's most numerous duck species.

4

5

and travel to a place of safety up-river. Caddis fly larvae (soft-bodied creatures with scant hope of survival on the turbulent stream bed) construct protective cases from whatever building materials are available, such as grains of sand, fragments of dead leaves and small shells. The stronger the current, the heavier the 'bricks'; sometimes even small pebbles are used. Other animals, such as freshwater mussels, burrow into the gravel of the stream bed or shelter under rocks in order to avoid the force of the current.

More commonly, however, the river denizens are too small to resist the current and must attach themselves to the bed by means of hooks, suckers or sticky slime. Riffle beetles attach themselves by means of hooks,   6

The cichlid fish *Haplochromis johnstoni* (6) is endemic to Lake Malawi in the East African Rift Valley. It is one of more than 600 species of cichlid found in fresh waters in Africa and South America, including the angelfish of the Amazon.

7

dragonfly larvae appendages are equipped with spines with which they cling to the substratum and blackfly larvae spin silken threads so that, should they be dislodged by the current, they can haul themselves back to safety. Freshwater leeches have suckers at both ends which they use to avoid being swept away, whilst the river limpet creates a vacuum beneath its shell by gripping hard with its powerful muscular foot, much in the same way as its namesake does on the seashore. In addition, research has shown that water molecules are so inhibited by friction that within a zone extending up to three millimetres from the river bed they scarcely move. Consequently this region is occupied by creatures such as the larvae of the delicate, ephemeral mayflies and larger stoneflies, which are flattened to reduce their water resistance.

As the river widens, the bed becomes more stable and accumulates sediments from upstream, enabling   8

Wetland mammals range from those with a seasonal interest in the water, such as the grizzly bear (7) in North America, which takes advantage of the salmon coming upriver to spawn, to the West Indian manatee or sea cow (2), a truly aquatic grazing mammal, distributed from the southeastern United States down to central Brazil.

Cutthroat trout (**4**) occur only in fresh waters on the western side of North America, while Atlantic salmon (**5**) breed in both European and North American rivers but spend their adult lives in the sea. The grey heron (**2**) is a predator of young fish and amphibians in both still and running waters.

Almost all of the eighty or so species of kingfisher are confined to the tropics and, despite their name, most are forest-dwelling species, their main prey being terrestrial creatures and airborne insects. African species include the lesser pied kingfisher (**1**), the woodland kingfisher (**6**) and the malachite kingfisher (**7**). Asian species include the Ceylon common kingfisher (**3**), a subspecies of the European common kingfisher (**12**), the rufous-backed (**8**), the white-breasted (**10,11**) and the blue-eared (**9**). The sacred kingfisher (**13**) of Australia will often nest in termite mounds.

certain pioneer plant species to take root. Many, such as water milfoils and river water dropworts, possess the characteristic dissected leaves which present least resistance to the current, while others have flexible, linear leaves, such as the unbranched bur-reed and some species of pondweed. Along the banks of these mature rivers, where the current is slower, luxuriant growths of arrow-head, water plantain, flowering rush and water speedwells may flourish beside dense stands of reedmace, bur reeds and numerous rushes, sedges and reeds. Less turbulent waters are also home to a greater variety of invertebrates, including a form of zooplankton comprising water-fleas and copepods, together with larger crustaceans such as freshwater shrimps and crayfish – bottom-dwellers that seek protection in the lee of boulders.

Typical freshwater fish include sticklebacks, bream, roach, tench and minnows in temperate rivers, which are preyed upon by such voracious predators as the pike. The salmon makes its way from the ocean to the upland river of its birth to spawn, negotiating waterfalls for distances of up to five miles en route. Likewise, the sturgeon spends its adult life at sea but lays its eggs in freshwater rivers and lakes. The human penchant for caviar – the eggs of this two-metre-long fish – has reduced the main breeding areas of the sturgeon in Europe to only a few sites, notably the French Gironde, the Guadalquivir River in southwest Spain and Lake Ladoga in northwest Russia.

Kingfishers, herons and dippers take advantage of the rich pickings to be had in the lower reaches of the river. The dipper is particularly adapted to its aquatic

habitat, feeding on invertebrates gleaned from between the pebbles on the river bed. In order to do so it 'flies' to the bottom of the river by flapping its wings and then proceeds to walk upstream, like a beachcomber battling his way against a storm. The cold waters of the high Andean rivers are populated by another truly aquatic bird known as the torrent duck. There are several species of this brightly-coloured bird, the distribution of which coincides almost exactly with the political divisions of the South American subcontinent. Like the dipper they face upstream and forage among the stones, using their webbed feet and stiff tails to maintain their position.

The temperate rivers of the world are also populated by several species of mammal. Some, such as the water

1

Typical mammals of flowing waters include the giant Brazilian otter (**3**), which is distinguished by its large size and flattened tail and may even feed on piranhas, and the American mink (**5**), a voracious predator whose diet includes fish, frogs, rodents and water birds.

2

vole of Europe and the North American muskrat, are herbivorous, others, such as water shrews and the Pyrenean desman, feed on small invertebrates. Both of the latter swim along the river bed against the current while seeking their prey with long, flexible muzzles. An inhabitant of mountains streams in northern Iberia and the Pyrenees, the desman is actually a rare member of the mole family, although it resembles a giant shrew. Other species, such as the otter and its close relative the mink, take larger prey. The otter, a fish-feeder in the main, is supremely suited to life in the water, having powerful webbed feet and a streamlined body to allow it to hunt its prey beneath the surface. Whilst under the water it can close its ears, and its pelt is remarkably thick and water-resistant.

Without a doubt the strangest river dweller outside the tropical region is the duck-billed platypus, a primitive, egg-laying mammal which is known only in freshwater lakes and rivers in eastern Australia. When scientists first examined skins of this beast they were convinced that they were hoaxes, contrived by some trickster

3

from the broad bill and webbed feet of a duck and the flat, naked tail of a beaver, all of which had been attached to a fur-covered body! Although this combination is less than ideal for travel on land, the duck-billed platypus is the epitome of elegance in the water. It forages for invertebrates on the river bed with its ears and eyes closed, locating its prey solely by means of tactile hairs on its head and leathery bill. This method enables the duck-billed platypus to hunt in even the murkiest waters.

For all their richness of animal life, the temperate rivers of the world cannot compete with the tropical waterways for sheer diversity of wildlife. The array of fish species here is vast, ranging from those that feed on aquatic vegetation or fruits and leaves which fall

4

5

6

7

8

9

10

African freshwater fish include the mouth-brooding cichlid *Melanochromis auratus* (4) of Lake Malawi and the primitive bichir (6) of the Congo which, like the African lungfish (8), is able to survive for short periods out of water, obtaining oxygen from the atmosphere through its lungs. A drying lake in Tanzania, however, spells doom for thousands of catfish (9), which do not have this ability. The voracious tiger fish (10) of the Nile, with its curved, dagger-like teeth, is the African counterpart of the South American piranha (7), one of the most predatory freshwater fish in the world.

from the surrounding trees, to voracious predators such as the piranha and the Nile perch. Fast-flowing waters house catfish and loach, which have either ventral fins converted into suckers with which they grip rocks, or else hang on with well-developed, fleshy lips. Typical bottom-dwelling fish of sediment-laden tropical rivers locate their prey using the sensitive barbels around their mouths.

Among the 1,300 fish species present in the Amazon is the arapaima, the largest freshwater fish in the world. It can reach three metres in length and weigh over 130 kilogrammes, but is a placid beast,

content to reside on the river bottom and feed on invertebrates and carrion. The coastal rivers to the north of the Amazon basin are home to the bizarre four-eyed fish, which has eyes divided into two sections: one adapted for underwater vision and one for keeping an eye on life above the surface. Amazonian hatchet fish take to the air to escape from aquatic hunters and, by flapping their pectoral fins, can cover distances of over five metres before returning to the water.

The most fearsome predator of all freshwater fish is the South American piranha, although even the largest species does not exceed sixty centimetres in

Great crested grebes (2) always breed on freshwater lakes, the adults carrying the young chicks on their backs to protect them from predators. Bank voles (1) are, in fact, more commonly found in deciduous woodlands and hedgerows than by rivers.

length. Despite their powerful jaws and formidable dentition, piranhas are rather unjustly regarded as bloodthirsty killers, attacking anything that moves. Although it is true that an animal the size of a capybara may be reduced to a skeleton in less than a minute by a shoal of hungry piranhas, these fish generally select only wounded or sick individuals, and in this way help to keep the rivers clean and free of disease. The electric eel is another predatory South American fish, often reaching two metres in length, although about eighty per cent of this is tail. It is the tail that encloses the eel's battery organs, of which there are two types: one is designed to help the eel find its way in murky waters by means of sonar – the eyes of the electric eel are so reduced that it is almost blind – and the other the eel

1

2

3

There are three families in the order Crocodilia: crocodiles and the false gavial, alligators and caimans, and the true gavial. In crocodiles, the fourth tooth of each half of the lower jaw can be seen when the mouth is closed, but in alligators these teeth do not show. In addition, crocodiles are sometimes found in brackish water and have even been known to swim out to sea, while alligators and gavials are strictly freshwater reptiles. When crocodilians are submerged only the eyes and nostrils show. To achieve this, some species even swallow stones to counterbalance the buoyancy of the lungs.

uses to stun its victims. Both head and tail must come into contact with its chosen prey simultaneously in order to complete the electrical circuit, but then the power so released can exceed 650 volts – sufficient to stun a horse.

The best-known tropical aquatic reptiles are the twenty-odd species of crocodiles and alligators, which have remained relatively unchanged over millennia and are the closest living relatives of dinosaurs. Unlike most reptile mothers, the females often stand guard over their nests of hard-shelled eggs, and in some species even assist the young in scrambling down to the water's edge. The young feed mainly on insect larvae and crustaceans, only changing to a fish and meat diet as they mature, and sometimes even adopting a cannibalistic habit and feeding on their siblings.

As crocodilians cannot chew their food, particularly large or tough creatures may be carried off to an underwater lair where they are secured under an overhanging ledge or submerged tree-trunk until tender enough to be chopped into pieces by these reptiles' sharp teeth. Their victims are usually land animals that have come to the water's edge to drink and are ambushed by crocodilians submerged in the river with only their eyes and nostrils protruding. Some species, however, favour a vegetable diet, such as the dwarf crocodile of West Africa, which has a predilection for fruit. The enormous, yet slender, Indian gavial, the

4

largest specimens of which commonly exceed seven metres in length, feeds on fish throughout its life. It differs from other crocodilians in having narrow jaws and a larger number of teeth, prefers deep, fast-running rivers and is probably the rarest of its kind, there being less than 300 remaining in the wild today.

Crocodiles and alligators are by no means the only reptilian predators of tropical rivers. The anaconda is probably the largest reptile in the world, its massive body stretching for up to nine metres. This member of the python family is almost exclusively aquatic, but will take up position in a waterside tree in order to ambush

Skull shape within the Crocodilia also varies, as can be seen in the narrower snout of the Nile crocodile (1,4,5) as compared to the broad outline of the American, or Mississippi, alligator (3).

5

6

Nile monitor lizards (**6,7**), some of which may be two metres long, have a predilection for the eggs of their neighbours, Nile crocodiles, although they also feed on a wide range of fish, aquatic rodents, birds and even freshwater turtles.

7

8

Caimans (**2,9**) are mainly South American alligators and may reach over seven metres in length. Similar in size is the gavial (**8**), sometimes known as the gharial, the only member of its family. In this species, whose distribution is restricted to the broad rivers of northern India, the slender snout may be up to four times as long as it is wide, and in male gavials the end of the snout bears a swollen protuberance. Unlike most crocodilians, which drown large animals, the gavial feeds almost exclusively on fish and has at least twenty-two teeth in each jaw, more than any other species.

9

land animals when they visit the river to drink. Like all pythons, the anaconda constricts its prey within its coils, rapidly suffocating its victim without even breaking bones, and then swallowing it whole. The metabolism of this snake is so slow that after one good meal it may not feed again for up to two and a half years.

Two particularly voracious turtles live alongside the anaconda in South American rivers, one of which is variously known as the tartaruga, giant arrau turtle or South American river turtle, the other as the matamata. The matamata is scarcely visible as it lies in ambush on the river bottom, its shell covered in algae and shreds of skin hanging from its head and neck. It appears to be in an advanced state of decay, but any animal foolish enough to venture within range is seized and then swallowed whole, since the matamata's weak jaws lack the necessary musculature for chewing. The tartaruga,

African fish eagles (**1**) are among the most characteristic birds of the Nile and East African lakes. They are highly territorial and pair for life, raising one or two chicks each year. Juvenile birds (**4**) at first lead a nomadic existence, becoming settled in one area at the age of four to five years.

probably the largest freshwater turtle in the world – the females may grow up to a metre long – is now extremely rare as its flesh is sought after as a delicacy. Most of the world's population is found in the sandy-banked rivers of the Orinoco basin in Venezuela and Colombia. The equivalent turtle in North American rivers is the alligator snapping turtle which, like the matamata, spends much of its life motionless on the river bed. Instead of relying on camouflage, however, it sits with its jaws wide open and attracts potential victims by twitching a small, red, wormlike lure at the back of its mouth.

Stretches of flowing water in tropical zones attract vast numbers of birds, especially where the river runs through an otherwise dry landscape. The Nile, for example, which traverses the Nubian and Libyan deserts as it flows towards the eastern Mediterranean, provides a valuable habitat for both invertebrates and fish-feeding birds. Along the bank strut black-headed and goliath herons, together with jabiru, or saddle-billed, storks, sacred and wood ibis and the bizarre whalehead stork. This latter, also known as the shoebill stork, stands over one-and-a-half metres high and has a massive bill with which it extracts lungfish from the drying mud at the edge of the river.

Birds which patrol the deeper waters in the centre of the river include pelicans, long-tailed cormorants, fish-eagles and ospreys, as well as a wide variety of

Nocturnal fish-eating birds include Pels fishing owl (**3**) in Africa and the brown fish owl (**8**), which has a range extending from the Mediterranean to China and wades into shallow water to snatch passing fish with its claws.

5

6

The marabou (**2**) is the largest member of the stork family, its massive bill being equally well suited to hooking sixty-centimetre lungfish from their refuges in drying lakes or plunging into the carcass of a wildebeest or zebra. Dabchicks (**9**) are the smallest European grebes, building their floating nests on freshwater lakes or along river banks.

8

7

9

African crowned cranes (**5**) are the most stunning members of the Gruidae, many of which stand over one-and-a-half metres high and have wingspans of up to two metres. Yellow-billed storks (**6**), also known as wood storks or wood ibis, now breed only in Florida in North America, despite being common further south. The sacred ibis (**7**), with its jet-black, naked head, is a gregarious bird of African wetlands south of the Sahara.

terns, skimmers and brightly-coloured kingfishers. Southeast Asia and Africa also boast fishing owls that have bare legs to help them to strike through the water cleanly and sharp-edged scales on their feet to provide a good purchase on their slippery prey. In the absence of such nocturnal predators in the Americas, the niche is filled by two species of fishing bat. Ranging from Mexico to the northern half of South America these bats have huge feet with sharp claws to secure their piscine victims, of which they may consume some thirty or forty in a single night.

Mammals which have learned to exploit the rich food supply of tropical rivers include those which feed on the bankside vegetation, such as the African water chevrotain, situtunga, puku antelope and lechwe, all of which have greatly elongated hooves, the toes connected by a flap of skin to support them on marshy ground. In South America, in the relative absence of ungulates, the riverside feeding niche is occupied by the capybara and tapir. Mammalian predators which take advantage of the wildlife which throngs to the river to drink and feed include such spotted cats as the ocelot, margay and jaguar in the New World, all of which are excellent swimmers and adept at catching fish. The jaguar is even capable of hauling a half-tonne manatee clear out of the water.

Pygmy Hippopotamus

☐ Distribution of the hippopotamus   ■ African swamp distribution

1

2

3

The Amazonian manatee is one of the few truly aquatic freshwater mammals, never leaving the water and using its prehensile lips to consume vast quantities of river plants such as the water hyacinth. Although an apparently cumbersome beast it is capable of reaching speeds of some twenty-five kilometres per hour, paddling its 'manhole cover' tail up and down to provide the impetus, and if necessary it can remain underwater for up to fifteen minutes. South American rivers are also home to several species of freshwater dolphin, including the inia, or Geoffrey's dolphin, of the Amazon and Orinoco systems and the La Plata dolphin which inhabits stretches of the Argentinian river of the same name, both of which differ from marine dolphins in that they have a clearly-defined neck. As the inia has spiritual significance for the tribal peoples of the Amazon basin it is not hunted, but the manatee – despite being the probable progenitor of the mermaid legends – does not enjoy such forbearance, and is widely trapped for its valuable flesh and fat.

Other freshwater dolphins are the Chinese river dolphin, or pei-chi, found in the Yangtse river, and the susu, which lives in the muddy waters of the Ganges,

4

Brahmaputra and Indus rivers of India. The susu differs from the other river dolphins in that it is completely blind – its eyes even lack lenses – and it navigates and locates its prey by sonar and with a highly sensitive snout and flippers.

The giant river otter can be distinguished from all other otters by its great size – adults may weigh more than thirty-five kilograms – and flattened, beaverlike tail. Its diet includes small water mammals and aquatic birds and their nestlings, but like all otters it favours fish, and will even feed on piranhas. In Africa, fast-flowing streams support populations of the Cape clawless otter, and there are spotted-necked otters to be found in slower waters. The capybara, the world's largest rodent, weighs seventy kilograms and stands sixty centimetres at the shoulder. It feeds largely in and

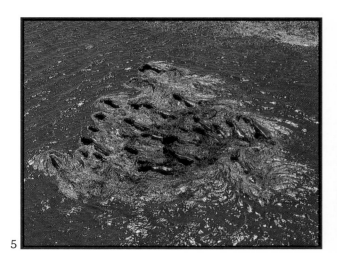

5

6

around the river on aquatic vegetation, but will retreat to the water if danger threatens. In undisturbed areas of northern South America and Panama, capybara are active mainly at dawn and dusk, but close to human habitation they have become nocturnal feeders.

A bull hippopotamus can weigh up to four-and-a-half tonnes and measure one-and-a-half metres at the shoulder, making it the third largest land mammal after the elephant and white rhinoceros. Although its name literally means 'river horse', the hippopotamus is actually a member of the pig family. Nature has endowed it with a barrel-shaped body, a broad muzzle and small eyes and ears, together with a thick, hairless skin. Hippopotamuses use the river during the day for a number of purposes. The water cools and lubricates their skins – a vital requirement as a hippo stranded away from the river in the heat of the African sun will gradually dehydrate; its skin will split and vultures may well arrive to feast on the animal even before it is dead. In addition, the water buoys up their massive bodies so that they use less energy in supporting themselves. At night the hippos leave the river to feed in the surrounding grasslands, returning to the water well before the sun comes up. They enrich the river by defecating in the water, thus transferring nutrients from the terrestrial to the aquatic habitat.

Although lakes are also freshwater habitats there are few species which are adapted to live in both running and still water. In fact, allowing for their lack of salinity, in their physical characteristics lakes bear

Herds of fifty to a hundred hippopotami (**3,4,5**) spend most of the day submerged in a river or lake, or wallowing in a muddy swamp, only leaving the water at night to feed (**2**) on the rich vegetation of the plains. Like those of the crocodile, the eyes, ears and nostrils of the hippopotamus are positioned on top of the head so that it can remain aware of its surroundings while submerged. The pygmy hippopotamus weighs, on average, only 250 kilogrammes, is a nocturnal, solitary beast and is far less dependent upon water than its larger counterpart, spending most of its time in the tropical forests of West Africa (**1**). The young of the manatee (**6**) are born under water. The female manatee's habit of adopting a vertical position whilst suckling her baby is probably the origin of the mermaid legend.

more resemblance to oceanic systems. The food chain is heavily dependent upon phytoplankton, and some of the deepest lakes have a distinct aphotic zone, where light never penetrates. Unlike oceans, however, which are continually stirred by currents, deep lakes are still and thus the rain of organic matter from the surface layers is not circulated, but is broken down relatively near the surface by bacteria which use up all the available oxygen very rapidly. These bacteria are replaced by anaerobic decomposers lower down, and the whole process results in a lack of oxygen which may limit the amount of life in deep lakes.

Rather simply, lakes can be regarded as islands in reverse, many of their inhabitants being restricted to one locality by their lack of dispersal mechanisms. Many adult invertebrates are capable of flight, by which means they can colonise nearby lakes, others have windblown egg capsules which are tolerant of desiccation, while many aquatic plants are able to spread to other areas through seeds transported in the mud adhering to the feet of waterfowl. Fish, however, are more or less confined to their lake of origin, although where the lake is drained by a river they may be swept downstream to a new location. On the whole,

1

2

3

4

The great Indian rhinoceros (**3**,**4**), largest of the three Asiatic species, today occurs mainly on the seasonally flooded plains of the Kaziranga nature reserve in Assam. Among defassa waterbuck, only the male (**2**) sports horns – handsome, ringed affairs up to a metre long. Black bears (**5**), known as cinnamon bears in the west of their range, are excellent swimmers and dextrous fishers.

lake denizens are ill-equipped to cope with the stresses of river life, and for this reason cannot colonise areas upstream.

The largest freshwater lake in the world, in terms of surface area, is Lake Superior (about 82,000 square kilometres), one of the five Great Lakes of the St Lawrence Seaway in North America which, like many lakes of temperate regions, are glacial in origin. Such glacial lakes are abundant across the whole of the northern hemisphere; Alaska, for example, boasts over three million glacial waterbodies more than twenty hectares in extent. Typically, glacial lakes are shallow, whereas those created by tectonic shifts in the earth's crust are usually deep and steep-sided. The East African Rift Valley, for example, was formed when movements of the adjoining continental plates in this region caused the subsidence of an elongated slab of rock, forming a narrow depression some 6,000 kilometres long and forty to sixty feet wide. The valley, which stretches from Turkey to Mozambique, was almost completely flooded at one time, but intense evaporation over millennia has confined the remaining lakes to a few of the lowest-lying areas.

Although known as 'freshwater' lakes, the majority

5

6

of the waterbodies in the Rift Valley are in fact salt lakes, due to the high levels of evaporation experienced in this area. By definition, salt lakes are those where dissolved minerals exceed thirty per cent, although unlike the oceans – where sodium chloride is the dominant component – Rift Valley lakes contain large quantities of sodium carbonate, and as such are also known as soda lakes. They are considerably more ancient than the glacially-formed waterbodies, being thought to be upwards of twenty million years old. By virtue of their great age and tropical location they are much richer in biological terms than the 12,000-year-old Great Lakes, which lie in a temperate region of the world. Lake Malawi, for example, possesses some 300-400 species of freshwater fish, the majority of which are not found elsewhere as a result of their long isolation from other waterbodies.

Lake Baikal, in the heart of the Soviet Union's coniferous forests, was also formed by a shifting of the earth's surface layers. Although covering an area of less than 31,000 square kilometres, it is immensely

8

Vertebrates that thrive in and around the world's wetlands range from the lugubrious toad *Bufo fowleri* (**7**) to the increasingly rare sandhill crane (**6**), the main population of which breeds in North America, in wetland areas near or within the Arctic Circle. The brightly-coloured but poisonous berries of the climbing plant woody nightshade (**8**) are a common sight in damp woodlands and marshes throughout Europe. The handsome Slavonian grebe (**1**) breeds on inland waters in northeastern Europe, just south of the tundra region, but winters in coastal regions further west and south.

7

deep – over one-and-a-half kilometres in places – and is thought to contain a fifth of all the world's fresh lake water: equivalent to that held by all five of the major Great Lakes put together. Baikal is also thought to be the oldest lake in the world, dating from some twenty-five million years ago. Thus, despite its location at about 60° N, it contains large numbers of species which are found nowhere else, most of which appear to be descended from creatures which once inhabited the headwaters of the rivers which feed it. Of Baikal's 1,200 species of animals and 500 plants, more than eighty per cent are endemic. Its most curious denizen is undoubtedly the Baikal seal, the only freshwater member of the group in the world, which is thought to have colonised the lake during the Ice Ages. Then frozen land supposedly made the journey from the sea relatively simple, even though Baikal is more than 2,000 kilometres inland.

Generally speaking, lakes may be divided into two categories depending on their nutrient status. Oligotrophic lakes are poor in nutrients and thus support only scant populations of aquatic organisms,

Both the Asian elephant (1) and the Indian, or water, buffalo (2) have been extensively domesticated in India and the surrounding region, largely as beasts of burden. As a result, truly wild animals are rather few and far between today.

1

2

whereas eutrophic lakes have dense phytoplankton and littoral vegetation and are also rich in animal life. Oligotrophic lakes are common in upland regions and are often of relatively recent origin, but eutrophic lakes are more usually encountered in the tropics and in temperate lowland areas where the quantity of nutrients received from rivers and the surrounding land is considerably higher. There is a gradual succession from oligotrophic to eutrophic conditions with age, although tropical lakes eutrophy more rapidly than temperate ones, largely due to the high levels of sunlight in the tropics, which encourage productivity at the lowest levels. This also applies to those which are shallower and overlay sedimentary rocks that provide more dissolved nutrients.

Around the shores of the lake lies the littoral zone, where the water is shallow and light penetrates to the lake floor. In temperate regions, bullrushes, reedmace and rooted floating plants such as waterlilies are common, accumulating sediments between their roots enabling them to gradually encroach on the open water. Consumer organisms are more prolific here than in any other part of the lake, and range from freshwater mussels, bottom-dwelling crayfish,

invertebrate nymphs and the herbivorous larvae of frogs, toads and newts, as well as the carnivorous adults.

Away from the banks lies the open-water zone, where the food chain depends almost entirely on phytoplankton – mostly diatoms and dinoflagellates – although some floating vascular plants such as duckweeds and frogbit may flourish if the water is unruffled. Large numbers of copepods and cladocerans make up the zooplankton, feeding primarily on the phytoplankton, and these in turn are eaten by a vast array of small fish species, including sticklebacks, minnows and bream. At the top of the food chain are predatory fish such as pike and piscatorial birds. Below the level of light penetration lies the plantless profundal zone inhabited by blood worms, which have such high levels of haemoglobin in their blood that they can thrive where oxygen levels are extremely low, alongside detritus-feeding bivalves and annelid worms.

So powerful is the attraction between water particles that the surface film of the lake is strong enough to support the weight of small invertebrates. Among the creatures that have evolved specifically to fill this niche are the predatory pond skaters, which have waxy,

The South American hoatzin (5) is sometimes considered to be the evolutionary link between the primitive fossil bird *Archaeopteryx* and modern-day species. This theory is based primarily on the fact that young hoatzins (7) possess two 'fingers', armed with strong claws, on the shoulder of each wing, although these are not present in the adult. Before they are able to fly, the young birds drop into the river when threatened, swimming (8) to safety and re-emerging only when the danger has passed.

3

The delicate African jacana, or lilytrotter (**4**), makes use of floating vegetation, especially lily leaves, to traverse the water in search of aquatic invertebrates. Tundra-breeding snow geese (**3**) gather in large flocks further south during the winter, often causing extensive damage to cereal crops.

4

5

6

7

8

(**6**) Water buffalo wallowing in a Bharatpur swamp provide a handy perch for cattle egrets. As the buffaloes move through the water, they disturb aquatic invertebrates and small vertebrates, which are then preyed on by these sharp-sighted herons.

hairy pads on their feet so as not to break the tension of the surface film. An airborne insect which has inadvertently flown too close to the water and become trapped by the surface film will create ripples in its struggles and so alert a pond skater. The predator will quickly remove its victim from the water's surface so that it will not have to share its meal with other skaters, then consume the body fluids by means of its long, piercing mouthparts.

The fish which occupy the various lakes of the world are no less diverse than those of rivers and sometimes, where they have been living in isolation for millennia, they are even more specialised. The elephant fish of the East African Rift Lakes, for example, has a greatly elongated face which forms a sensory 'trunk' with which it rummages for food on the lake bottom. It is also capable of producing an electrical discharge, although weak indeed compared to that of the electric eel, which it uses for navigation, especially in murky waters. The electric catfish is another creature which can generate its own electricity, this time given out as

1

2

3

Lesser flamingos (**2**,**3**), less than a metre in height, are abundant in Africa, some three million breeding in the Rift Valley alone, while greater flamingos (**5**), standing up to 127 centimetres tall, also range over Europe, Asia and America, but are less common in Africa. Because of their different feeding habits, the two species are not in competition for food. Lesser flamingos prefer algae and feed in the more saline shallows, whilst greater flamingos feed on aquatic arthropods in the deeper waters. Flamingos often coexist with pelicans (**1**) as the latter exploit yet another food resource – fish.

a powerful jolt of 350 volts or more, with which it can both defend itself from its enemies and stun smaller fish for food. The related upside-down catfish is even more specialised, feeding near the surface, but swimming belly-up at all times.

Although many amphibians, especially toads, spend much of their adult lives on land, the biology of this group obliges many of them to return to the water to reproduce. In temperate regions, spring sees the familiar strings and globular masses of spawn in lakes, ponds and even puddles. Among the most graceful amphibians are the newts, and none more so than the great crested newt of Eurasia, the males having a bright orange belly and a spiky crest running along the length of their body in the breeding season. More bizarre, however, is the newt-like axolotl, actually a salamander, which is found only in Mexico's Lake Xochimilco. It is thought that a lack of iodine in the waters of this lake prevents the axolotl from attaining its adult form, but even though it must remain in the water, and still bears the gills of the larval form, its reproductive organs eventually become fully developed. It is, in effect, a sexually-mature tadpole – a state of affairs known as neoteny. Neoteny is also seen in another salamander, the mudpuppy of North American rivers and ponds, which is so-named because it is believed to bark like a dog.

4

The albino axolotl (**7**) is the most famous of the neotenous salamanders – those that can reproduce without metamorphosing from the juvenile state to the adult. The Argentinian horned frog (**4**) may reach twenty centimetres in length, sufficient to dispatch a large rat, while midwife toads (**8**), at the opposite end of the size scale, are renowned for their habit of carrying their eggs around until they hatch.

5

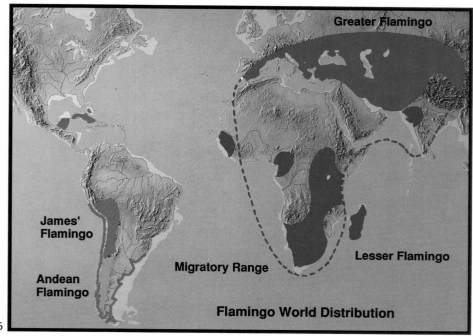

Greater Flamingo

James' Flamingo

Andean Flamingo

Migratory Range

Lesser Flamingo

Flamingo World Distribution

6

7

8

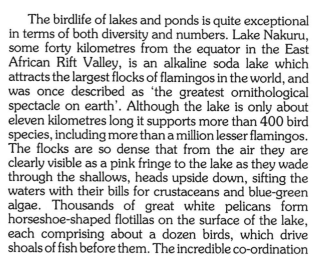

9

10

The birdlife of lakes and ponds is quite exceptional in terms of both diversity and numbers. Lake Nakuru, some forty kilometres from the equator in the East African Rift Valley, is an alkaline soda lake which attracts the largest flocks of flamingos in the world, and was once described as 'the greatest ornithological spectacle on earth'. Although the lake is only about eleven kilometres long it supports more than 400 bird species, including more than a million lesser flamingos. The flocks are so dense that from the air they are clearly visible as a pink fringe to the lake as they wade through the shallows, heads upside down, sifting the waters with their bills for crustaceans and blue-green algae. Thousands of great white pelicans form horseshoe-shaped flotillas on the surface of the lake, each comprising about a dozen birds, which drive shoals of fish before them. The incredible co-ordination

The North American Plains spadefoot (**9**) is named for the horny 'spades' on its hind feet, used for digging. Daubenton's bat (**10**) is the only British species to be seen regularly hawking for gnats and other aquatic winged insects over water from dusk onwards.

of the birds is exemplified by the fact that they dip their beaks into the water in unison every ten seconds or so. During the winter months, the presence of tens of thousands of Eurasian ducks and waders, including teal, shoveller, pintail, stints, sandpipers and ruffs, feeding alongside African spoonbills and little egrets, is clear evidence of the value of the Rift Valley as a migration route for birds wintering in the south.

The fact that so many waterfowl can exist side by side on the world's lakes and marshes is largely due to their diverse feeding techniques. Sawbills, such as red-breasted mergansers, smew and goosanders, dive for fish, trapping them securely with narrow, serrated mandibles, while the so-called diving ducks – red-crested pochard, tufted ducks, scaup, ferruginous ducks and goldeneye, among others – swim to the bottom to feed on aquatic plants and invertebrates. The dabbling ducks, including the ubiquitous mallard, as well as pintails, gadwall, wigeon, teal, garganey and North American black ducks, generally filter detritus and plankton from the water between their flattened mandibles, frequently upending in order to do so. The largest waterfowl are the swans, found on all continents except Africa, which use their long necks to reach down to the lake bottom to feed on deep-growing plants.

The most completely aquatic of all freshwater birds are undoubtedly the grebes, which rarely leave the water, spending most of their lives on one lake or pond. This predilection for the water has led to reduced

1

2

wings in many species, and some, such as the giant pied-billed grebe of Lake Atitlán in Guatemala, have lost the power of flight completely. In all grebes the feet are placed well to the rear of the body to provide maximum impulsion, and the strong toes are partially-webbed as, like sawbills, these birds dive to obtain their food, mostly small fish and aquatic invertebrates. Darters, also known as anhingas or snake-birds, are also remarkably adapted for an aquatic existence, being able to expel the air trapped in their feathers in order to swim beneath the surface to catch fish. Their underwater prowess is facilitated by strong, webbed feet and the fact that they also flap their wings under water to provide momentum.

4

Around 5,000 species of dragonfly (**1**) frequent the world's freshwater wetlands, most of which lie in tropical regions. The catchfly ragged robin (**3**) is an attractive plant of European marshlands. Typical wetland birds include the Eurasian snipe (**7**), here incubating its newly-hatched young, the western grebe (**5**) of North America and the spoonbill (**8**), closely related to ibises, of southern Europe and North Africa. The South American capybara (**4**), a denizen of flooded grasslands and river margins, is the world's largest rodent, while the Canadian beaver (**10**) creates its own lake by constructing a dam out of trees felled with its impressive incisors.

There is one freshwater animal that actually creates lakes in the course of its everyday life: the beaver. There are two very closely related species of beaver, differing primarily in their geographic distribution: the Canadian beaver occurs in North America, while the European species is confined to Eurasia. Like grebes and otters, beavers are supremely adapted to an aquatic way of life, having broad, webbed hind feet and a flat, oarlike tail for propulsion, dense, water-repellent fur, and transparent eyelids, nostrils and ears which close whilst it is swimming under water. Beavers are the second largest of the world's rodents after the capybara, weighing up to eighteen kilogrammes at maturity. Their strong teeth continue to grow throughout their lives, being constantly re-sharpened as the beaver gnaws its way through trees up to a metre in diameter.

Beavers normally live in temperate rivers in forested regions. By gnawing through the trunks of the riverside trees they fell large amounts of timber which they then use to construct a dam across the river. This floods a large area, creating a deep lake, the main purpose of

5

6

7

8

9

10

Characteristic animals of Sri Lanka's wetlands include the spotted deer, or chital (**6**), the pond heron (**2**), also known as the paddy bird as it is commonly seen in rice fields, and the pheasant-tailed jacana (**9**), which has similar habits to its African relatives.

which is to enable the beavers to venture further afield in search of trees, the branches and foliage of which also provide their main foodstuffs. Somewhere in the middle of this lake the beaver family will build its lodge, completely surrounded by water and thus well out of reach of terrestrial predators. Even at the height of winter the beavers will have access to the lodge via one or maybe two underwater entrances, since the depths of the lake will rarely freeze.

Swamps and marshes are generally formed by the gradual infilling of shallow lakes, although some tropical wetlands may occur in river deltas. The Sudd swamplands of the Sudan, for example, are dominated by stands of herbaceous papyrus and have a permanent area of almost 100,000 square kilometres, which increases sevenfold in the rainy season. In contrast, Florida's Everglades – at over 30,000 square kilometres, the largest freshwater swamp in the United States – is covered for the most part with trees, particularly mangroves by the coast, and bald cypresses inland. A large proportion of the world's 8,800 known bird species choose to live in and around such wetlands, particularly those of the heron family, which number sixty species, and kingfishers, of which there are some ninety species.

The crane family, many members of which are very rare today, is particularly dependent upon a rich marshland habitat. The Siberian crane breeds in the tundra but winters in Asian wetlands such as that of Bharatpur in India. This beautiful bird numbered thousands during the nineteenth century but, following

Freshwater invertebrates range from the hot springs spider (5), found in Yellowstone National Park, to more attractive species such as the tiger swallowtail butterfly (9) and the European damselfly *Agrion virgo* (2), characteristic of clear, fast-flowing streams. The bodies of the females are always green, the males generally being blue.

1

3

4

5

6

Although freshwater wetlands (7) are among the most productive ecosystems in Europe in terms of natural vegetation, they are unsuitable for agriculture and have therefore suffered from widespread drainage in recent decades. Among the more typical acid bog plants are the sundews (1,7), whose gland-tipped hairs trap and digest small insects to provide the plant with nitrates not readily available in these habitats. Other European wetland plants include gipsywort (4), a member of the mint family, and whorled lousewort (8), an attractive, semi-parasitic species of montane flushes.

European Bogs,
Marshes and
Wetlands

the loss of its wintering wetlands and severe hunting in India, there are barely hundreds in the world now. On the other side of the world, the North American whooping crane was once common in the marshlands of that continent, but by 1954 had been reduced to only twenty-one individuals. This situation is now being remedied by a captive breeding programme.

A chapter on freshwater wetlands would not be complete without at least a mention of peatlands, of which it is estimated that there are only about 200,000 square kilometres – a little over 0.1% of the total land area of the earth. The extent of peat in the British Isles is some 15,000 square kilometres and includes the single largest extent of peat in the northern hemisphere

– that of the Flow Country of Caithness and Sutherland in Scotland. Peatlands are generally restricted to oceanic coasts with high rainfall between latitudes of 45° and 65°, the only inland area being on the slopes of the mist-drenched Ruwenzori Mountains in Central Africa. Notable peatlands exist in the Aleutian Islands, the Norwegian Highlands, Iceland and the Gulf of St Lawrence in the northern hemisphere, while, south of the equator, there are some on New Zealand's South Island, the Falkland Islands and Tierra del Fuego.

Peatlands represent some of the most desolate of all habitats, formed over millennia in conditions where waterlogged soils prevent the rapid decomposition of dead plant material, which subsequently builds up as

North American whooping cranes (**6**) are among the world's most endangered birds, there being only a few dozen left in existence – the basis of a captive breeding programme. African hammerkops (**3**) mate for life, constructing an enormous nest some two metres in diameter that is added to each year.

White storks (**3**) breed across Europe, central Asia and Africa south of the Sahara, wintering a little to the south of this range. They are as likely to nest on elevated buildings as in trees, occupying the same site year after year.

All sixty members of the heron family (**2**) have long, pointed bills, ideal for spearing fish or amphibians, long legs for wading through the water in search of prey, and webbed toes to give some support on the soft muds. The forty-five-centimetre-high North American green heron (**4**) has shorter than average legs and can therefore wade through shallow water only; unlike most herons it sometimes watches from a perch and dives into the water to catch its prey. The nyala (**5**) is a small, southern African antelope that is rarely found far from water.

peat. The dominant plant of these ecosystems is a small moss known as sphagnum, of which there are numerous species, each suited to a particular set of environmental conditions. The remaining flora is equally highly specialised in order to exist in such a habitat. Cotton-grasses, rushes and sedges are abundant, as is cross-leaved heath, which grows with cranberry, bog rosemary and bog pimpernel, and such small shrubs as bog myrtle in Europe and Labrador tea in North America. Dragonflies, moths and diving beetles are among the most common animal species of peatlands, and some of them are so well-suited to these habitats that they are unable to exist elsewhere.

Most interesting of all, however, are the many plants which have evolved methods of supplementing their nutrient intake by becoming carnivorous. Sundews, of which there are about ninety species in the world, mostly in Europe, are predominantly peatland plants. Their leaves are clothed in sticky glandular hairs which trap unwary insects, secrete enzymes to break down the corpses and absorb the nutrients so released. In North America, pitcher plants produce cylindrical leaves at ground level which are half-filled with a liquid

6

8

Various species of sugar cane (**7**) are found in the wild state from Africa to southeast Asia. Despite reaching over three metres in height, sugar cane is a type of grass, cultivated mainly in the tropics as the world's most important source of sugar.

7

A huge diversity of wetland birds is able to coexist because of the variety of their feeding techniques and preferred prey. The beak of the skimmer (**6**) has a keeled lower mandible that is superbly adapted to scooping small aquatic creatures from the water without the bird missing a wingbeat. Grey-headed gulls (**8**) more usually feed from the surface, or scavenge at the margins of the lake. The bizarre shoebill stork (**1**), also known as the whalehead, has a beak so massive that the young chicks are unable to lift their heads. The hefty, angled bill is ideal, however, for extracting lungfish from their cocoons in the drying mud of lakes in the Sudan, Uganda and the southeast Congo region, such victims being located by the bird's long-toed feet. Even newly-hatched crocodiles and turtles provide food for this solitary, territorial member of the stork family.

containing digestive enzymes. Insects are attracted to the plant by nectar-secreting organs lying just inside the pitcher. Having entered, they are unable to climb out because of the waxy nature of the inner surface, sometimes supplemented by downwards-pointing teeth on the rim.

Raised bogs are thought to take between seven and eight thousand years to reach a height of ten metres, yet the exploitation of the peatland resource far exceeds the rate at which it can regenerate. Dried peat, consisting as it does of compressed plant material, burns at a very high temperature and for a long time. Peat-cutting by the local people for fuel has been in progress for hundreds of years in Ireland as trees are scarce over much of the country, but this century has seen the appearance of huge, peat-fired power stations and distilleries, with machines resembling combine harvesters being used to remove peat from the bogs.

Unfortunately, peatlands are not the only threatened wetlands in the world today. Reclamation for agriculture is destroying thousands of hectares of marshland in temperate regions, whilst the drainage of wetlands to get rid of disease-bearing mosquitos is devastating

tropical and subtropical swamps and lakes. Elsewhere, irrigation of dry croplands is crucially lowering the water table by depleting the underground aquifers in surrounding areas; wetlands such as Spain's Tablas de Daimiel have almost completely dried out as a result. Rivers are suffering from pollution, both by contaminants and by heat, and in many regions they are canalised between concrete embankments, eliminating the littoral communities they otherwise might have supported.

Even recreational activities such as water-skiing, sailing and wind-surfing can result in the disturbance of breeding waterfowl, whilst the wash from powerboats devastates waterside vegetation and contributes to the erosion of the banks themselves. In the last decade yet another threat has reared its head: acid rain. It is thought that airborne pollutants in car exhaust fumes and power station emissions descend to earth with rain and snow and are accumulating in lakes and rivers, acidifying their waters and killing whole communities of aquatic organisms. It is possible that some badly-hit lakes, especially in Scandinavia and Scotland, will never recover.

· CHAPTER 10 ·
TROPICAL FORESTS

Despite occupying only about six per cent of the earth's land surface, tropical forests support a profusion of animal and plant life that is unrivalled anywhere else in the world. The enormous buttressed trees and the delicate epiphytic orchids and ferns which festoon their boughs; the millions of insect species, many of which have yet to be discovered by man; the magnificent jungle tiger and man's closest living relatives – the anthropoid apes – are all part of this rich and varied belt of equatorial forest. And yet man, whose ancestors emerged from the jungles of Africa only a few million years ago, is destroying this wealth at an ever-increasing rate. We must bear in mind that all those creatures which vanish in the name of human progress are gone forever.

Right: the lordly tiger, (top) multicoloured rainbow parakeets and (above) an Asian siamang.

The tropical forest, also known as the equatorial forest or jungle, is one of the world's least extensive habitat types, covering only about six per cent of the total land area of the earth. Its nine million square kilometres are distributed over three main areas, all of which lie within a narrow equatorial belt extending to about fifteen degrees north and south of the equator. Central Africa accounts for about nineteen per cent of the total forestation, most of which lies in the Congo basin, though there is a further area stretching along the coastal lowlands of West Africa. Southeast Asia and Oceania account for a further twenty-one per cent of the world's tropical forest, about half of which lies in Indonesia. Latin America that contains about fifty-five per cent of the total, most of which lies in the Amazon

basin, centred on Brazil. There are also tracts of tropical forest in Queensland, northwest Australia, and on the islands of New Guinea and Madagascar.

Only decades ago it was estimated that the total number of living species in the world numbered about five million, but recent research suggests that there may be over thirty million species of insect alone. Regardless of the size of this figure, it is thought that almost half the world's species are found in tropical forests, and man has probably yet to discover many of them. Despite their limited extent, there is no terrestrial habitat to equal the tropical forests for sheer diversity and abundance of wildlife, both animal and vegetable. Only tropical coral reefs can rival them.

The total number of plant species in Amazonia has been estimated at more than 40,000. About 25,000 species of flowering plant are thought to exist in Southeast Asian equatorial forests. In tropical forests generally, the number of tree species per hectare ranges from fifty to two hundred, as compared to only a handful of species in the temperate deciduous forests of Europe. Moreover, one hectare of South American jungle supports up to 40,000 invertebrate species. Indonesia, most of which is covered with tropical forests, contains seventeen per cent of the world's bird species and more than a hundred endemic mammals, while more than one fifth of all bird species live in Amazonia. But despite this wealth, the distribution of tropical forest animals usually follows one of two patterns: either they are confined to a very restricted area, sometimes only one tiny valley; or very low densities of individuals are scattered over a much larger area. Thus each patch of forest harbours a highly individual assemblage of animals, forming a unique and quite irreplaceable ecosystem.

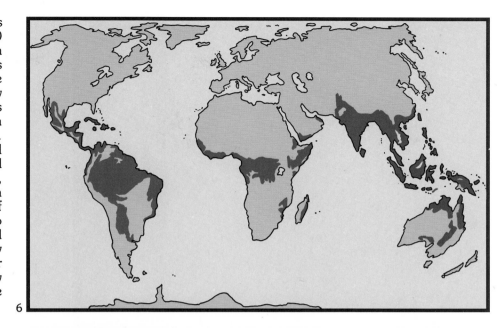

6

7

The distribution of tropical forests (6) across the face of the globe is distinctly equatorial. These forests have been the source of many plants of economic value, including the banana (2), the largest herb in the world, wild species of which are native to Malaysia and Indonesia. The nipa palm (3) of Southeast Asia is a sugar-producing member of the olive family and is one of the first terrestrial species to invade mangrove swamps. The climbing or creeping glory lilies, including the magnificent *Gloriosa superba* (5), are native to tropical Africa and Asia. The common chameleon (7) has two pairs of opposing toes on each foot, as does the Australian rainbow parakeet (1), while the Costa Rican tree frog *Hyla ebloceatta* (4) has suction pads at the tips of its toes. Such adaptations assist these creatures in their arboreal lifestyle.

This cornucopia of wildlife is thought to owe its existence both to the extreme age of the forests, which amounts to tens of millions of years during which species evolved to fill all available niches, and to the stability of the mature jungle, preserved by constant climatic conditions. The equatorial location of these forests ensures that sunlight, day length and temperature are the same all year round, and that there are only minor variations in the level of rainfall over the course of twelve months. The two most important factors are temperature and rainfall: the former is around twenty-seven degrees Celcius, the latter over 2,000 millimetres annually. More importantly, the total precipitation is more or less evenly distributed throughout the year. The combination of intense rainfall and high average temperatures, exacerbated by forest evaporation and transpiration, creates conditions of extreme humidity at all times; in the region of ninety-five to ninety-nine per cent.

As a result of variation in rainfall, several different types of tropical forest can be distinguished. The type that grows in the wettest conditions, often under an annual precipitation of over 4,000 millimetres, is called, appropriately, tropical moist forest or tropical rainforest. The majority of trees are evergreen, shedding their leaves a few at a time throughout the year. Rainforest forms the heartland of the world's tropical forests and accounts for some two thirds of the total. Amazonia, the Congo, Sumatra, Borneo and Papua New Guinea are rainforest areas. A little further from the equator, where rainfall is lower – between 2,000 and 4,000 millimetres per year – an element of seasonality creeps in. The lush, evergreen jungles are replaced by tropical semi-deciduous forests, also known as tropical seasonal forests.

The majority of scorpions (2) live either in deserts or tropical regions. They use their powerful pincers to dismember their prey, only applying the venomous sting if the victims struggle. Caterpillars such as those of the *Thosea* butterfly (1) often have spiny appendages to break up their outlines and so camouflage them.

Typical of this latter form, in which the driest month produces less than fifty millimetres of rain, are the monsoon forests of Burma and Thailand; subtropical forests in Central America and the Caribbean; the West African forests, and those in the northeast corner of Australia. The dominant canopy trees are deciduous, shedding their leaves and becoming dormant for a short period each year when rain is scarce, but the understorey shrubs and small trees are mainly evergreen. As a rule, semideciduous forests are less species-rich than rainforests and, because of the build-up of dead leaves on the forest floor during the dry season, fire is an important factor in their growth and regeneration. Marginal tropical forests in some regions, especially Southeast Asia, may be almost completely composed of deciduous trees.

Tropical forests have been in existence since Tertiary times, if not before. Over the ensuing sixty-five million years, highly efficient nutrient and mineral cycling systems have evolved within them. Characteristically, rainforest soils are thin and poor in nutrients. They can support such luxuriant plant growth purely because seventy per cent of all mineral elements are held within the vegetation itself at any one time. Owing to high temperatures and an abundance of moisture all year round, biological activity is rapid. Fallen leaves, branches and fruit are rapidly broken down into their component molecules by a vast array of decomposer microorganisms, and the nutrients so released are immediately absorbed by the dense mat of roots which lies beneath the soil surface. Many forest trees have symbiotic mycorrhizal fungi and bacteria living in association with their roots that greatly increase the efficiency with which they can take up mineral salts from the soil.

The 'Tarzan' image of the jungle – impenetrable foliage, snakes dripping from the trees, a multitude of huge, exotic flowers and hourly confrontations with dangerous wild animals – is, on the whole, misleading. Tropical rainforests have often been likened to

6

There are eight distinct races of tiger (**4,7**), three of which have become extinct in recent years due to human pressures of their environment. The stripes, which provide excellent camouflage among the shadows of the tropical forest, are never symmetrical, and are more prominent in the island races.

7

cathedrals: a twilight world of enormous, fluted columns rising from wide, open floors, for there is little in the way of vegetation at ground level to impede the progress of either man or beast. The cacophony of howls and screams that forms the soundtrack of many a jungle film occurs mainly at dawn and dusk, and the owners of these disembodied voices are rarely seen. Many live their whole lives high in the trees, without ever setting foot on the ground. Only where the canopy is more open, permitting sunlight to penetrate to the ground, does the undergrowth become dense. This occurs, for example, along river banks and roadways, in gaps left by fallen trees, or where the virgin forest has been cleared and replaced by a tangled mass of secondary vegetation. Monsoon forests, by virtue of a predominance of deciduous trees, which allow the undergrowth access to light at certain times of year, tend to support more vegetation at ground level than rainforests.

Seen from the air, the tallest trees of tropical forest, some exceeding fifty metres in height, emerge from

the canopy as from a frothy green sea. They are widely spaced and, as a consequence, have room to develop a spreading, umbrella-shaped crown. However, the canopy itself is composed of densely packed trees with compact, rounded crowns which absorb seventy to eighty per cent of all sunlight falling on the forest. The trunks are normally straight, branchless and many times longer than the depth of the foliage. Beneath the canopy, which is normally twenty-five to thirty-five metres above ground level, lesser trees grow. Their crowns are much narrower in proportion to their height. Closer to the ground, however, the vegetation is sparse, as little sunlight penetrates through the triple tree layer above. A few shrubs may overshadow the field layer, which consists almost entirely of plants which do not rely on photosynthesis to grow and reproduce. Instead they obtain their nutrients saprophytically with the aid of fungi, or live parasitically on the roots of trees.

Viewed from the forest floor, many of the trees look remarkably similar to each other, having thin, smooth

Asian elephants (**5**) are smaller than their African counterparts as size is no great advantage in a densely-forested environment. The woolly opossum (**6**) of Central and South America is a small arboreal marsupial with long, thick fur. American alligators (**3**) are typical inhabitants of the lush swamplands of the southern United States, such as Florida's Everglades region.

barks, since there is no risk of either desiccation or fire in these humid conditions. The largest of the trees support their tremendous height by means of plank buttresses, which may spread many metres away from the relatively slender main trunk. Those leaves which can be examined from the floor are evergreen and elongated, with smooth, waxy upper surfaces, rather like laurel leaves, narrowing to a 'drip-tip'. The function of both wax and drip-tip is to shed water rapidly after rainfall in order to prevent the growth of epiphytic mosses and fungi, which would restrict the leaf's ability to absorb light. By disposing of any excess water on the leaf, these features also renew the processes of evaporation and transpiration from the leaves, thus ensuring that a continuous supply of water, carrying mineral salts essential for growth, is drawn up from the soil via the roots.

The uniformity of leaf shapes and the lack of distinguishing bark features belies the diversity of rainforest trees. Only when they flower is it possible to distinguish between the many different species. The lack of seasonality in the forest creates an apparently haphazard flowering cycle, although it is thought that the rhythm unique to each species may be dependent

upon some internal timing mechanism. Some will bloom almost continuously throughout the year, and some produce flowers at more or less annual intervals. Others may flower only once a decade, or even less frequently. Since all members of a single species will flower at the same time, it is possible to distinguish another feature of the tropical forest – the widely spaced distribution of trees of the same species. Rarely are such individuals found adjacent to one another, which is far from the case in temperate forests, where single-species stands over large areas are common.

The lack of air movement within a tropical forest, due to the dense foliage, means that almost all flowers produced beneath the canopy are pollinated by animals and their seeds dispersed by the same animals. Only emergent trees have access to a breeze and, therefore, often bear inconspicuous, wind-pollinated flowers and produce light, fluffy or winged seeds which are carried away on the air currents. The animal pollinators of the lower levels are mainly winged insects and birds which

Nipa palm inflorescences (**1**) produce large quantities of nectar that attract pollinating bees and also yield a commercially valuable syrup. The flowers and fruit of the cacao tree (**2**), a native of the eastern equatorial slopes of the Andes that is widely cultivated in tropical Africa and America today, emerge straight from the trunk, the pods being the source of cocoa beans. Trailing, epiphytic lianas (**3**) are one of the most characteristic sights of the tropical forest. The cannonball tree (**4**) of tropical America has showy flowers and tough spherical fruits from which it derives its name.

5

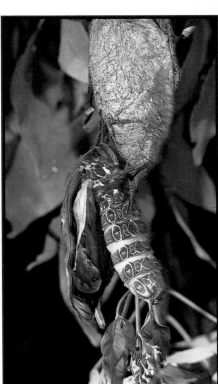

6

7

can travel with ease from flower to flower. On some of the smaller trees, the flowers sprout directly from the trunks in order not to be obscured by the canopy vegetation. This also enables larger creatures to play the part of pollinator creatures such as lemurs in Madagascar, and their African relatives, bushbabies. Some pale, night-opening blooms, often emitting a foetid odour, are even designed to entice bats as pollinators.

Most tropical forest trees produce large seeds containing a rich supply of nutrients to give the seedling a good start in life regardless of the impoverished forest soils. Those which are dispersed by animals are attractive, succulent fruits, often brightly coloured to attract attention. The animal dines well in return for depositing the undigested seeds elsewhere in its faeces. The fact that at any one time only a few tree species will be producing fruit means that animals have to travel over long distances to find food. The dispersal of the seeds will thus be widespread, continuing the

trend of mixed stands which is so characteristic of tropical forests.

In the relative absence of light at the lowest levels, seedlings of rainforest trees grow very slowly, adopting a state of near dormancy once they have attained a height of a few metres and sometimes remaining that way for decades. But lightning, violent winds, or the weight of water produced by a sudden downpour may prove too much for an ancient canopy tree, or one of the emergent giants, and it will fall, often taking a number of the surrounding trees with it. This creates a considerable gap in the canopy, through which sunlight streams unimpeded, and the seedlings grab their chance. In a rapid burst of growth the young trees compete vigorously for the advantages inherent in a place in the canopy. Only a few will succeed, the others succumbing in the struggle.

The lack of light on the forest floor is undoubtedly the reason why epiphytic plants are so abundant in tropical forests. The sunlit upper branches are festooned

The giant Atlas moth of southeast Asian rainforests is one of the world's largest insects. The enormous caterpillars (**5**) spin silken cocoons from which the adults emerge with crumpled wings that soon expand to their full twenty-five-centimetre span (**6,7**).

with all manner of botanical hitchhikers, most of which produce aerial roots to absorb moisture from the humid forest air and trap decaying organic matter between stems and leaves to supply their mineral needs. Others, such as members of the mistletoe family, are less innocuous, penetrating the bark of their host tree to obtain their nutrients. The commonest epiphytic plants are those with light seeds which drift through the canopy on the breeze, or those bearing seeds thar are dispersed in fruit consumed by birds. Their sticky seeds adhere to the branch where they are regurgitated or deposited in the birds faeces. Bromeliads and orchids are especially abundant, as are ferns, mosses, liverworts and lichens. In the Amazon rainforest there is even an epiphytic species of cactus.

Such climbers belong to a wide range of plant families, generally represented in tropical forests by woody lianas. Liana seedlings germinate on the forest floor and need little light, but the mature plants require high intensity sunlight. So, as they develop, lianas attach themselves to surrounding trees and bushes by means of tendrils, suckers and hooks, gradually hauling their stems upwards. Some lianas may be as thick as a man's thigh, but others are much more slender, such as the rotang palm – from which rattan is derived – of which there are around a hundred species in the Old World. Lianas bind together the various components of the forest, so that a falling giant will, as a result of these ties with adjacent trees, create a much larger gap in the canopy than the one it previously occupied.

The Sabah pitcher plant (**2**) is one of a number of carnivorous Old World species, usually epiphytic vines, in which the tendrils expand to form flasks half-filled with a sugary liquid to attract and trap insects and so provide the plant with additional nitrogen.

1

2

Another epiphyte, known as the strangler fig, produces aerial roots that, rather than absorbing atmospheric moisture, continue to grow until they have reached the forest floor. From there they supply the plant with water and mineral salts. During the lifetime of the plant these roots gradually thicken and finally envelop the host tree completely so that, denied access to light and air, it dies. By this time, however, the strangler's roots are so thick and solid that they are capable of supporting their own weight, and a freestanding 'tree' is the result.

Other plants gain sufficient light by exploiting existing vegetation in order to reach higher levels.

3

Birdwing butterflies, such as Rajah Brooke's birdwing (**1**) of Malaysia, live in forested habitats from southern India to Australia. The males are splendid creatures with iridescent markings and are often attracted to puddles on the forest floor, while the dull females prefer the canopy. Sunbirds are basically an African group although some, such as the purple sunbird (**3**), are found in Asia. They are the Old World equivalent of the hummingbirds, hovering in front of flowers to extract nectar with their long, curved beaks and tubular tongues. Like the birdwings, only the males bear resplendent metallic plumage.

The dim world of the forest floor is not a lifeless place, despite the lack of light. Fungi sprout from rotting wood in a multitude of shapes and sizes, their colours ranging from vivid purples to almost luminous scarlets and oranges. Fungi lack chlorophyll and therefore do not photosynthesise like higher plants. Instead they secrete enzymes which break down dead leaves, wood and animal matter into their component chemicals. These are then absorbed by the dense network of fungal mycelium – thread-like structures which represent the body of the fungus beneath the soil. When sufficient energy reserves have been built up, the fruiting bodies are produced: familiar toadstools and mushrooms, among others. The abundant moisture and high temperature of tropical forests favour the

The night-blooming *Sonneratia* (**8**) is one of the first mangrove species to colonise silt-accumulating sections of the Indonesian coast, while the attractive pink-red inflorescences of the torch ginger (**6**) are found in the inland forests of these tropical islands.

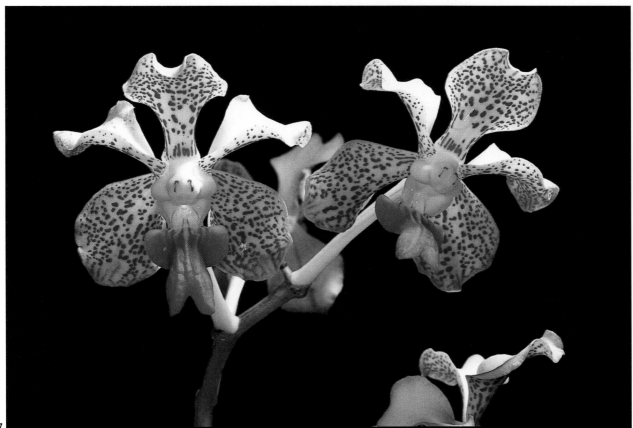

The Orchidaceae is the second-largest family in the plant kingdom, its members numbering between 15,000 and 35,000 species. They are found on every continent except Antarctica, but the majority occur as epiphytes in tropical-forest habitats, such as the quarterterete vanda (**4**) and *Vanda tricolor* (**7**) of southeast Asia. The genus *Dendrobium* numbers over 1,000 species distributed from the lower slopes of the Himalayas to southern Australia, one of the more spectacular being the Indonesian *Dendrobium stratiotes* (**5**).

fungi and, in return, the nutrients contained within dead trees are quickly returned to the system.

Just as the plant life of the tropical forest is divided into distinct levels, so are its animal inhabitants. The emergent trees are colonised mostly by birds and insects, whereas the forest floor supports ground-dwelling deer, rodents, peccaries and wild pigs. But it is the region inbetween that holds the greatest variety of animal life. This ranges from tree-dwelling monkeys to sloths, anteaters and small, arboreal carnivores – many of which never descend to the ground. The dense foliage limits the usefulness of sight, and the lack of even the lightest breeze beneath the canopy renders the sense of smell virtually useless. As a consequence, most vertebrates have highly developed hearing. Camouflage is important as a means of avoiding predators, and is also employed by the hunters themselves, especially those that prefer to lie in wait for their victims.

Among the smaller forest creatures, many moths and their larvae blend in with the foliage of their favourite food plants, and sap-sucking treehoppers clustered on a branch resemble thorns. The iridescent blues and greens of the *Morpho* butterflies decorate the upper surface of the wings only, so that when the insects settle and fold their wings, the drab colouring of the underside renders them almost invisible. Praying mantises secrete themselves among brightly coloured flowers, or wait motionless among the leaves. Their coloration and shape depend upon their choice of hiding place. Predatory spiders hide among the petals of the forest flowers. Some are even able to change colour according to the shade of the bloom, an ability also displayed by chameleons. Many insects have large, eye-like patterns on their wings which, if opened suddenly, shock a potential predator into thinking that its intended victim is in fact a much larger creature; this strategy is fairly common in butterflies and crickets.

The lack of vegetation at ground level in a tropical forest precludes the presence of such huge numbers of large herbivores as are found in the surrounding grasslands – the East African savannas or the llanos of Venezuela, for example. Nevertheless, the clearings created by fallen trees support a much richer growth of herbaceous plants, and it is here that many herbivores emerging from the shelter of the surrounding forest find their food. Asiatic elephants and the great Indian, Javan and Sumatran rhinoceroses are all smaller and better adapted to the forest than their plains-dwelling

1

3

2

4

5

Jungle invertebrates often attain a prodigious size, such as the two-horned rhinoceros beetle (**2**) of Java, the long-nosed weevil (**3**), and the giant millipede (**5**) of Costa Rica. Katydids (**4**) have leaflike bodies which serve to camouflage them from predators. Fungi, such as this delicate South American species (**1**), flourish in the damp litter of the forest floor.

African relatives. All three rhino species are now endangered. There are less than fifty Javan rhinos living in the wild today, and the Sumatran species is only better placed, numbering around 150 individuals.

The Asiatic elephant, sometimes called the Indian elephant, differs from the African species in a number of ways. It is essentially a jungle animal and has smaller ears, only one prehensile 'finger' at the tip of its trunk, four nails on its hind feet (the African elephant has only three) and a rounded, convex back. Unlike the African elephant, the Asiatic species is easily domesticated, and has, accordingly, played a prominent part in the history of this part of the world as a beast of burden, used particularly in timber extraction and hunting. Most Asiatic elephants inhabit the tropical forests of Assam today and, like the forest rhinoceroses, are endangered in their natural habitat.

6

The Indian, or one-horned, rhinoceros (**10**) is today found mainly in the marshlands surrounding the Brahmaputra River. The related Javan rhinoceros (**6**) is the world's rarest pachyderm, now confined to the dense forests of the Udjung Kulon reserve in western Java, despite being quite common across Southeast Asia in the nineteenth century.

7

8

9

Female Asian elephants have very small tusks, only a few centimetres long, but those of the male often exceed a metre in length. Their sight and hearing are poor, but their sense of smell is very acute, enabling them both to locate lush vegetation and to detect potential danger (**7**,**8**,**9**).

10

11

The South American harpy eagle (**11**) is the largest and possibly the rarest of the New World eagles, living in dense tropical jungles and distinguished by the two-pronged crest in the adult.

The gaur (**7**), the largest species of wild cattle found in the jungles of Southeast Asia, stands almost two metres high at the shoulder and weighs up to a tonne. The smaller banteng (**5**) is found in the wild only in Burma, Vietnam, Java and Borneo, having been widely domesticated elsewhere.

Large herbivores of the lower forest are predominantly browsers rather than grazers, feeding on the leaves and shoots of small shrubs since grasses are scarce. The only true African forest browsers are the bongo – among the largest of all African antelopes – and the secretive okapi, a relative of the giraffe. The large Asian rainforest herbivores include such rare bovine species as the gaur and the banteng, (both of which have been successfully domesticated) as well as the endangered kouprey, which was not discovered until 1937 and is thought to be the ancestor of the zebu cattle of India. These wild forest cattle are disappearing mainly because of interbreeding with domesticated strains and the loss of their habitat. Asian forests are also home to many members of the deer family: the axis, or Indian spotted deer; muntjaks, or barking deer; Timor deer and the Indian sambar. The Bornean wild boar and the Indian, or crested, wild boar also frequent the lower levels of the forest, as do the white-lipped and collared peccaries in South America.

Smaller creatures are represented by the mouse deer of Asia, which is, despite its name, a member of the pig family, and the diminutive African duikers and royal antelopes – the latter are usually only forty centimetres long from head to tail. The rodents which fill this niche in South America include capybaras, coypus, pacas and agoutis. Most of these are not true browsers, feeding on a variety of vegetable matter, including leaves and shoots, fallen fruit, seeds, nuts and

Asian elephants (**1**) must be frequent visitors to water, the adults drinking an average of 230 litres daily. Rhinoceros horn is composed of compacted keratin fibres like hair and is regarded as a powerful aphrodisiac in the Far East. As a result, both Javan (**3**) and Indian (**4**) rhinoceroses have suffered at the hands of poachers and their numbers have been decimated this century. The timid mouse-deer, or chevrotain (**6**), stands less than thirty centimetres high and is found in swampy forest habitats in southeast Asia. The Cape pangolin (**8**) is one of four African species and is about one metre long, with a scaly coat, powerful claws to rip open termite mounds and an extensile tongue to scoop up the insects.

fungi, and usually venturing out into the forest clearings at night. Some will even eat insects, fish and carrion.

Ground-dwelling birds are usually bulky creatures and seldom take to the air. Characteristic species are the pheasants in Southeast Asia and the turkey-like curassows in South America, the Malaysian jungle fowl (believed to be the ancestor of the domestic chicken) and the Congo peafowl, the only African member of the peacock family, which was unknown to science until 1936. In the tropical forests of Australia and New Guinea, the flightless cassowary is the largest ground-dwelling bird. It defends itself against enemies and members of its own species which invade its territory by kicking out at the intruder with its large feet, which are armed with spikes.

But all ants have their enemies. Pangolins in Africa and Asia, and the tamanduas and two-toed anteaters of South America – creatures closely related to the giant anteater of the plains – descend to the ground to feed on them, usually breaking open termite nests and ant colonies with their strong claws and consuming the frantic inhabitants with the aid of their long, sticky tongues. Other creatures take advantage of the moving columns of ants, rather than attacking them in their nests. These include the South American ant birds, of which there are more than 220 species.

However, few birds inhabit the lower levels of the tropical forest compared to the numbers which live in the canopy and above, although a visitor to the forest floor will catch only frustrating glimpses of these birds

Cassowaries (**2**) are large, flightless Australasian birds with strong legs and a horny casque on their foreheads. Peacocks (**9**) are found in Asian and African equatorial forests. The males are renowned for their magnificent 'eyed' tail feathers, which open like a fan when displayed.

6

7

8

Other birds spend much of their lives among the trees, descending to the forest floor only at the beginning of the breeding season. The South American cocks-of-the-rock, and many of the bowerbirds and birds of paradise of northeast Australia and New Guinea, conduct their courtship displays in clearings where the magnificent plumage of the males can be shown off to best advantage. Typically, the male performs a series of intricate dance steps, usually associated with much spreading of feathers. The healthiest males will undoubtedly be in possession of the most splendid plumage and dance most vigorously, thus attracting a larger number of females. This selection process ensures that only high quality genetic material is passed on from generation to generation.

Typical ground-dwelling insects are forest ants, some of which, such as termites, leaf-cutter and harvester ants, are peaceable creatures. However, others, especially the army ants, carve a trail of destruction across the forest floor. They consume any small creatures, including the nestlings of ground-dwelling birds, which may lie in their path. These ants, described by some as the most terrible predators of the South American rainforest, differ from their relatives in that they are primarily nomadic. Their columns may number up to 15,000 individuals and, as they are blind, they use scent trails in order to maintain formation.

9

silhouetted against the tree-lined roof. The tiny, iridescent hummingbirds are confined to the New World and, although they range across a wide variety of habitats, the majority are to be found in the tropical forests. The smallest is the Cuban bee hummingbird, measuring less than five centimetres from beak to tail. Even the largest, the giant hummingbird, is only about twenty centimetres long. None of the 300 or so species is at home on the ground because they have weak feet and toes. For this reason they spend much of their lives on the wing, possessing better developed flight muscles than any other bird in proportion to their size. Those species which dine predominantly on insects have short, pointed bills, whilst nectar-feeders have long bills, often curving downwards, and tubular tongues.

In every respect hummingbirds live life at high speed. They beat their wings at up to eighty strokes per second, enabling them to fly at around 100 kilometres per hour. Their rapid wingbeat gives them great manoeuvrability. Many species hover in front of flowers to feed, and some can even fly backwards should the need arise. One consequence of such an accelerated existence, however, is a high metabolic rate – a typical hummingbird's pulse will beat 1,260 times per minute, and it will breathe around 270 times in the same period – which means that it must feed almost continuously during daylight hours in order to obtain sufficient energy to maintain its body processes. At night it will become torpid, reducing its body temperature and metabolism almost to the point of hibernation, thus saving valuable energy.

1

2

3

Most bats are nocturnal, such as the Mexican *Chrotopterus* (**3**), locating their airborne insect prey and avoiding obstacles by echolocation, though some, like the fruit-eating flying foxes (**5**), do rely more on scent and vision. Most bats are also dark-coloured, exceptions including the white Costa Rican leaf-folding bats (**4**).

4

Another denizen of South American tropical forests is the hoatzin, thought to be the closest living decendent of the extinct genus *Archaeopteryx* – the earliest bird genus to appear in fossil evidence. The main reason for this line of thought is that, like *Archaeopteryx*, young hoatzins are born with a pair of strong-clawed fingers at the tip of each wing to help them clamber through the trees. These fingers disappear after a few weeks, but the adult birds still attempt to use their wings to catch hold of branches, often damaging their flight feathers in the process. Unlike hummingbirds, hoatzins are poor fliers, owing to the lack of development of their chest muscles, and the chicks will escape predators by dropping into the river and swimming away rather than by taking flight.

Noisy and gregarious, never descending to the ground, toucans are close relatives of the woodpecker and are endowed with oversized, saw-toothed bills. Although their plumage is at best funereal, their bills are brightly coloured. Bill patterns are relied upon to distinguish species in the field. The bulky bill is packed with air-filled cells, and so is not as cumbersome as it might at first appear. However, it does make feeding awkward – the toucan must toss the insect, small vertebrate or piece of fruit into the air with its bill, tipping its head right back to receive the morsel.

The Old World ecological equivalent of the hummingbirds are the sunbirds of Africa and Asia, while toucans are replaced in Africa by hornbills. The beak of some species of hornbill supports a large horny

The African and Asian hornbills are characterised by a hollow, bony casque that lies above the already massive bill. It is better developed in the male than the female, as can be seen by comparing the female wreathed hornbill (**7**) of Indonesia with the male Malabar pied hornbill (**9**).

5

casque. The function of this structure is unclear, although it may amplify the voice and allow the hornbill to be heard over long distances, which could prove invaluable for locating a mate in the dense forest. Touracos are also found in Old World tropical forests. They are agile, tree-dwelling birds, possessing an outer toe which can be directed either forwards or backwards, proving useful in grasping branches or food.

Another predominantly airborne tropical forest creature is the bat. There are more species of bat in the South American rainforest than in any other part of the world. Some forest species, such as the South American ghost and proboscis bats, are insectivorous. Others are nectar feeders, pollinating the flowers they visit in the process, whilst some of the world's largest bats are

There are more than 300 species of hummingbird, most of which live in tropical forests. These iridescent birds, close relatives of swifts, have well-developed flight muscles that enable them to hover in order to extract nectar from flowers with their bills and to fly at speeds of almost a hundred kilometres per hour. Their bills vary considerably in shape, being almost needlelike in such species as the sword-billed hummingbird (**1**). The female is responsible for nest construction (**6**), laying a pair of eggs two or three times per year, and the young (**2**) are remarkable in that they do not develop the downy plumage of most chicks, instead sprouting adult feathers straight from the naked skin.

7

6

8

fruit-feeders, such as the African hammer-headed fruit bat, which has a wingspan of almost a metre, and the so-called flying foxes of Asian rainforests. Other species are purely carnivorous –the bulldog bat will eat only fish, whilst others take rodents, fledgling birds and even other bats. The vampire bats of South America feed exclusively on the blood of large mammals, making a small incision in a hairless part of their body with their sharp teeth. The amount of blood they consume is unlikely to affect their victims seriously, but there is a high risk of disease being transmitted in this way.

The forest habitat has probably encouraged an evolution of flight in a variety of animals which are earthbound in other, treeless habitats. Not having to descend to the forest floor in order to search for new

9

Like hornbills, toucans (**8**) have large, brightly-coloured beaks filled with air cells and probably designed to be a warning to other species. These South American tropical forest birds are gregarious and noisy and may measure some sixty centimetres from head to tail.

feeding grounds has obvious energy-saving advantages, and flight is also useful in escaping from potential predators. This evolutionary phenomenon has reached its fullest development in the jungles of the Indo-Malayan peninsular which, as a result, has often been referred to as the 'realm of flying animals'.

In fact, most of these so-called 'flying' beasts are actually gliders, having developed a variety of mechanisms whereby the surface area of the body can be increased to provide greater air resistance. The two species of flying lemur, or colugo, for example, have developed flaps of skin which extend from the tip of the tail to the neck, creating a hexagonal 'parachute', or patagium, which enables them to glide for over fifty metres between trees, losing only ten to fifteen metres in height. Despite their name, these creatures are not true lemurs but are members of an ancient, squirrel-like family which has no close living relatives.

Other gliding creatures found in the tropical forests of Asia are the flying squirrels, which have developed furry membranes between their hind and forelimbs in much the same way as flying lemurs. They use their longer tails for steering. The five species of giant flying squirrel have so perfected this technique that they can glide for over 450 metres in one leap, though smaller species are limited to flights of thirty to forty metres. Flying lizards also possess flaps of skin along their sides, stiffened by bony projections of the ribs. Flying

Like all anteaters, the terrestrial giant anteater (**1**) has a keen sense of smell, strong claws for ripping open termite mounds and rotting logs, and a long, sticky tongue. Its smaller relatives, the tamandua (**6**) and the two-toed, or little, anteater (**7**), are true, arboreal forest animals, with long, prehensile tails and shorter snouts. Two-toed sloths, or unaus (**2**), are larger, lighter-coloured and faster-moving than their three-toed relatives (**4**), also known as ais; both can turn their heads through almost 360°. Asian pangolins (**8**) are distinguished from their African cousins by the presence of external ears. All three species are arboreal.

1

2

3

4

5

Most lorikeets are nectar-feeders with bristled tongues and play an important role in pollination (**3**). Barbets, named for the long bristles which extend from the bill, are mostly African species, except for the Asian genus *Megalaima*, which includes the blue-throated barbet (**5**).

frogs and geckos adopt a slightly different approach to aerial travel. Instead of extensions of skin from the body, they have enormously long toes joined together by thin membranes, providing them with four separate 'parachutes'. The paradise flying snake of Borneo employs a different technique again. It can be up to two metres long and is able to glide for about fifty metres. When it takes to the air this snake flattens its ribcage to produce a ribbon-like profile, creating at the same time a ventral cavity, and arranges its body into a series of 'S'-shaped coils, all of which help to increase air resistance.

However, there are many mammals occupying the upper layers of the tropical forest that have no such

6

American alligators (**9**) feed mainly on vertebrate prey, especially slow-moving fish and terrestrial animals that come to the water's edge to drink. Small creatures are usually swallowed whole, but larger prey is often stored underwater, to tenderise it.

7

8

gliding abilities. Instead, adaptations for an arboreal existence include a prehensile tail, as in New World monkeys, pangolins and tamanduas and a few carnivores, such as the kinkajou and binturong. This tail acts almost as a fifth limb, like the opposable thumbs of the primates and sharp claws of many other mammals, which enable their owners to grasp branches more firmly. Binocular vision, a further adaptation, is essential to the faster-moving arboreal creatures for judging distances between branches. Rodents, especially squirrels, are also well-represented in the forest canopy. They range in size from the diminutive striped palm squirrel to the giant Malabar squirrel of Southeast Asia, which is almost one metre long. Other Asian rodents include the tree rats, replaced in South America by a multitude of spiny rats, ranging from seven to fifty centimetres in length and inhabiting all levels of the forest.

The phlegmatic sloths of South America can be divided into the slightly less torpid two-toed species, or unaus, characterised by brownish fur and relatively

9

large eyes; and the three-toed sloths, or ais, which have green-tinged fur owing to growths of algae. These are more common in the lower trees. Three-toed sloths spend most of their lives suspended upside down, their well-developed claws hooked over the branch, and their limbs stiffened like coat-hangers to support the burden. In order to shed the rain, the hair on the flanks and legs of these inverted creatures grows in the reverse direction: that is, from ventral to dorsal. All sloths are leaf-eaters, although two-toed species consume fruit as well, and have extremely long and efficient digestive systems in order to cope with large quantities of tough foliage.

1

4

2

Other arboreal inhabitants of tropical forests around the world are the primates, the majority of which spend their lives high in the canopy, and rarely descend to the ground.

Occupying an intermediate position between insectivores and primates, and classified with each group alternately according to varying scientific opinion, are the thirty-odd species of Asian tree shrews. Slightly more advanced are the huge-eyed slow, slender and pygmy lorises, which prowl the Asian jungle at night. But even closer to true monkeys is the tiny creature known as the tarsier, which measures only about thirty centimtres in length, two-thirds of which consists of tail. There are three species of tarsier, all restricted to a few islands in Indonesia and the Philippines. They are

3

5

6

Tropical South American tree-dwellers range from the marsupial woolly opossum (**8**) to the coatimundi (**1**), a relative of the panda noted for its intelligence. Nocturnal, vegetarian tree porcupines (**6**) have quills which assist more in camouflage than defence. All of these creatures feature on the menu of the harpy eagle (**3**), nicknamed the 'leopard of the air'.

7

almost frog-like in appearance, and have well-developed hind legs, enabling them to travel up to five metres in a single jump, and suckers on the tips of their fingers and toes to ensure that they land safely. Like the loris, the tarsier is a nocturnal creature with enormous eyes, and its highly flexible neck enables it to turn its head through a full 180 degrees.

Such primitive primates also occur in Africa. Bushbabies, or galagos, are nocturnal and extremely agile, escaping danger by leaping from branch to branch. However, the more lethargic potto never jumps. If threatened it will curl itself into a ball, revealing spine-like extensions of the cervical vertebrae, although these are not adequate defence against predators with a strong set of teeth. Yet another prosimian group, the lemurs, were isolated from their mainland relatives on the island of Madagascar some fifty million years ago and, as a result, they exhibit a great diversity of anatomy and lifestyle. The many species of giant lemur are now extinct, and although such curious beasts as the aye-aye, indri and sifaka still exist today, most are either rare or endangered owing to the degradation of their jungle habitat.

Australasia has no indigenous primate species whatsoever. As a result, its rainforests are strangely silent. In Australia, this niche is filled instead by arboreal marsupials such as the phalangers, or possums, and tree kangaroos. As a result of convergent evolution, possums have many features in common with the primitive primates, including stereoscopic vision, opposable thumbs and prehensile tails. The cuscus is one of the largest possums, a slow-moving creature that resembles the South American sloths in many respects. Other possums bear a strong resemblance to

8

Siamangs (**2**), denizens of Malaya and Sumatra, are related to gibbons but distinguished from them by the membranes linking the second and third digits and a laryngeal sac which swells to amplify calls. The Indian grey langur (**7**) is a slender, agile monkey that is exclusively vegetarian and has long intestines to cope with this cellulose diet. Abyssinian guerezas, or colobus monkeys (**4**), are found in family groups high in the canopy of African tropical forests and are very agile, despite lacking thumbs. The Madagascan sifaka (**5**), although monkey-like, is in fact a lemur, feeding on leaves, shoots and fruit, which it grasps directly in its mouth rather than feeding itself using its hands.

the lemurs. The striped possum, for example, is the Australasian counterpart of the aye-aye, whilst another species is even called the lemuroid possum, so similar is it to the Madagascan prosimians. Other phalangers have independently evolved the ability to glide from tree to tree, especially in New Guinea, and the marsupial sugar glider is very similar in appearance to the flying squirrel.

True monkeys belong to five families: *Cebidae* (howler and spider monkeys); *Callitrichidae* (tamarins and marmosets); *Cercopithecidae* (by far the largest group, including mangabeys, guenons, colobus monkeys, langurs, baboons, mandrills, macaques and snub-nosed monkeys); *Hylobatidae* (gibbons); and *Pongidae* (the great apes - gorillas, chimpanzees and

grammes, is probably the pygmy marmoset, which vocalises at ultrasonic frequencies.

Monkeys belonging to the family *Cercopithecidea* are found in both Asian and African tropical forests. These Old World monkeys are characterised by narrow noses and a remarkable repertoire of facial and vocal patterns for recognition purposes. They occupy all levels of the forest, from the floor to the emergent trees. In Africa, for example, Diana and blue monkeys are found in the tallest trees; acrobatic, leaf-eating colobus monkeys, or guerezas, in the canopy, and black-cheeked white-nosed monkeys spend the night safely hidden in the canopy, descending to forest clearings to feed during the day. The smallest Old World monkey, the talapoin (less than forty centimetres

Langurs, also known as leaf monkeys, are sacred in India and are thus a common sight around towns and villages as well as in their more natural forest domain. Both grey langurs (**1**) and capped langurs (**3**), typical species of the Assam region, form small, well-organised social groups.

1

orang-utans). The first two families occur only in the New World. They are distinguished from the Old World monkeys by their broad noses with widely spaced nostrils, and their less-opposable thumbs. Many of the American species also have prehensile tails, and it is not uncommon for the male to take part in rearing the young. Both these attributes are lacking in African and Asian monkeys.

There are over sixty primate species in Central and South America, more than two thirds of which are found in Brazil, including such endangered monkeys as the three types of lion tamarins – golden, golden-headed and golden-rumped. Several New World primates are endemic to Brazilian Amazonia, including the tassel-eared marmoset, bald uakari, black saki and long-haired spider monkey. The douroucouli, or night monkey, of South American rainforests is the only truly nocturnal monkey species in the world. Its enormous eyes give it an owl-like appearance. The commonest New World monkeys are the capuchins, thought to rival chimpanzees in intelligence. The smallest monkey, weighing only about eighty-five

2

Siamangs (**2**), unlike most Asian primates, are equally at home in both lowland and montane tropical forests – they have been recorded living at altitudes above 3,000 metres. They have the long arms typical of the gibbon family, moving through the trees by brachiation. Macaques, also found across Asia, are almost exclusively terrestrial primates with omnivorous feeding habits. They are perhaps better known as rhesus monkeys and are used extensively in medical research, with the consequence that numbers are dwindling in the wild. Long-tailed macaques (**4**), found in Malaysia, maintain a rigid class structure within the troop.

long), occupies the understorey shrub layer, along with Hamlyn's and Brazza's monkeys. Some species, such as mandrills and mangabeys, are more at home on the ground.

Asian representatives of this family include the langurs, or leaf monkeys, which occupy much the same niche as the African guerezas, and the macaques, or rhesus monkeys, which, like mandrills, have well-developed fangs. The most abundant forest macaques include the common rhesus monkey, also known as the Indian macaque, lion-tailed macaques and toque and bonnet monkeys – all of which live in large groups and have a well-defined social hierarchy. Rhesus monkeys are, in fact, fairly close relatives of the grassland baboons, also exhibiting a predominantly terrestrial lifestyle. Although regarded as sacred across much of Asia, they are widely used in medical research, and hundreds of thousands are collected and sold for vivisection purposes every year. The leaf-eating langurs, with the typical long intestine and large stomach of herbivorous species, are also sacred in India. They include such common species as the purple-faced and douc langurs and the dusky leaf monkey.

Because of their resemblance to man, gibbons and pongids are collectively known as the anthropoid apes. Those which inhabit Asia – the various species of gibbon and the orang-utan – are all arboreal, whereas the two African species – gorillas and chimpanzees – are equally at home on the forest floor. The gibbons are **7**

Most of South America's primates are confined to the rainforest environment, particularly that of the Amazon Basin. They include the tamarins, such as the distinctive cotton-top (**5**), a small, agile omnivore whose tail makes up a quarter of its body length. In contrast, the bare-faced uakari (**6**) has a stumpy tail and moves more slowly through the forest, rarely leaping from tree to tree and seldom descending to ground level. The black-headed spider monkey (**7**) of Central America has long, slender limbs and a tail with enormous gripping power that serves almost as a fifth limb.

the smallest anthropoid apes. They are supremely adapted to an arboreal way of life in that they have narrow, long-fingered hands and long arms, the shoulder joint of which is mobile through 180 degrees in any direction, enabling them to swing through the trees by a process known as brachiation. The most widely distributed is the lar gibbon, but there are many other species with a much more restricted range. The wau wau, for example, is confined to Java and Borneo, the dark-handed gibbon occurs only on Sumatra; the black gibbon is found in Vietnam, northern Laos and the island of Hainan and the two species of siamang are restricted to Malaya and Sumatra.

Another Asian anthropoid ape, the orang-utan, is found only on the islands of Sumatra and Borneo. An adult male stands around one and a half metres tall and may weigh up to ninety kilogrammes, the females being smaller. Like gibbons, they move by brachiation.

1

The terrestrial anthropoid apes include the chimpanzee (**4**) and gorilla (**3,5**) of Africa's equatorial forests, while the tree-dwelling orang-utan (**1,6,7**) is confined to virgin rainforest in the Malayan archipelago. Both eastern subspecies of gorilla are larger and hairier than the western race, and the adult males have a powdering of white hairs on the upper back.

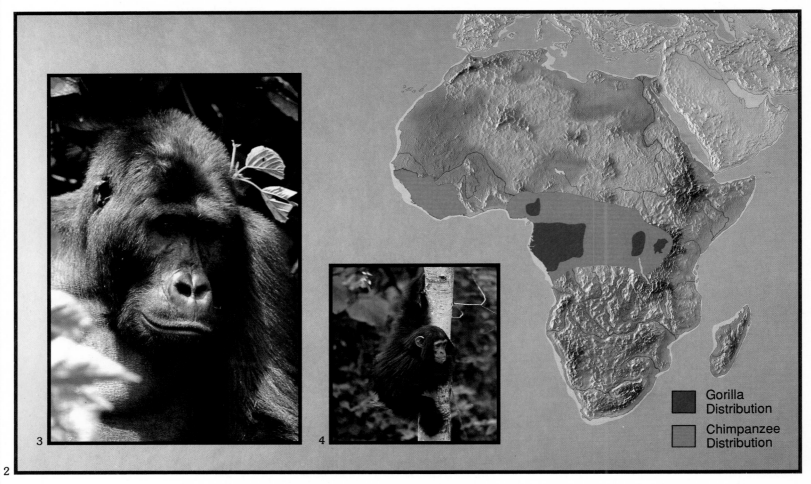

3

4

Gorilla Distribution

Chimpanzee Distribution

2

The absence of leopards on these islands means that orang-utans have very few natural predators and, without a doubt, man is their greatest enemy today. The Sumatran population crashed from around 1,000 individuals in 1963 to only about a hundred by the following year, largely due to live poaching to supply the zoo trade. Stricter controls in this international trade have since permitted numbers to recover somewhat.

The African gorillas are similarly threatened by the activities of man, although, in the case of the endangered mountain gorilla, which is now confined to its last bastion in the Ruwenzori range, gorilla poaching is concerned more with the collection of trophies, especially heads, than with live capture. The eastern lowland gorilla, and its counterpart further west in equatorial Africa, are slightly better off in terms of total population size, although they too are threatened with habitat loss through the encroachment of agricultural lands on the jungle, and by illegal hunting. Although gorillas are primarily terrestrial apes, chimpanzees divide their time between the canopy and the forest floor, according to the available food supply.

5

Venomous snakes are found throughout the world's rainforests. They range from South American pit vipers, such as the fer-de-lance and the bushmaster, which have small depressions between their eyes and nostrils to sense temperature changes, and thus the presence of warm-blooded creatures, to African rhinoceros and Gaboon vipers, boomslangs and the fast-moving black and green mambas, which can reach speeds of up to twenty-five kilometres per hour along the ground. Pythons are mainly arboreal snakes, ambushing their prey from the branches of a tree. The largest in the world, often up to eight metres long, is the reticulated python of Indonesia and the Malay peninsula. The related Indian python is a mere six metres long. Instead of injecting venom into their prey, which consists of birds and small mammals, they suffocate it by throwing their considerable length around it in coils and then swallow it whole.

6

7

8

The orang-utan is traditionally known as the 'man of the woods' on account of its almost human appearance. The males stand up to one-and-a-half metres tall and weigh up to ninety kilogrammes, while the females are considerably smaller. Both sexes develop layers of subcutaneous fat under the naked facial skin as they get older. Rarely descending to the ground, orang-utans lead a solitary life, moving through the trees by brachiation and sleeping in specially-constructed platforms high in the canopy. When a female grey langur (8) gives birth to her single baby she leaves the troop in order to devote more time to her offspring. After some fifteen months the youngster must fend for itself when she mates again.

Each of the three main regions of tropical forest boasts one large raptor which terrorises the upper levels of the jungle. In Africa, this is the crowned hawk eagle; in Southeast Asia, the Philippine monkey-eating eagle, and in South America, the harpy eagle. These birds, all of which stand over a metre high, have a fondness for arboreal mammals, especially monkeys. They have short, rounded wings and long tails for silent manoeuvrability through the trees, keen vision and hearing to locate their prey, and huge talons and a powerful, hooked bill with which to seize and tear apart their victims.

Other rainforest raptors seek smaller prey. The Congo serpent eagle feeds mainly on small, arboreal snakes; the ornate hawk eagle prefers birds; Ayre's hawk eagle feeds on smaller mammals such as squirrels and will also take birds, and the bat hawk hunts around rivers at dusk. The scavengers of the South American rainforest include turkey and yellow-headed vultures which, unlike their plains-dwelling relatives, have a highly developed sense of smell in order to pinpoint freshly killed carcases. The dense tree cover prevents them from being able to locate their prey visually.

Mammalian carnivores vary greatly in size and diet and, as a result, tropical forests support more such species than any other habitat. Members of the mongoose family are omnivorous, especially the nocturnal, arboreal civets, which are found in both

1

2

South American rainforest raptors include the handsome nocturnal spectacled owl (**1**), a particular enemy of the little anteater, and the magnificent ornate hawk eagle (**2**), whose powerful claws are employed more frequently in seizing large birds such as parrots and curassows than mammals.

3

4

5

Most tropical forest frogs are small, brightly-coloured and arboreal. The Latin American family Dendrobatidae includes the arrow-poison frogs, the most poisonous in the world, including the one-centimetre-long poison dart frog *Phyllobates lugubris* (**8,9**), the red and blue poison arrow frog *Dendrobates pumilio* (**11**) and other *Dendrobates* species (**6,10**), many of which show a high degree of parental care. Other New World species include the harlequin frog *Atelopus varius* (**12**), more closely related, in fact, to the toads, and also highly toxic, the golden toad *Bufo periglenes* (**7**), which, unlike most arboreal frogs, has a very weak call, and the tree frog *Hyla rosenbergi* (**13**), with suction pads at the ends of its digits.

African and Asian jungles, although far more species inhabit the latter, including linsangs, palm and Malay civets and binturongs. The binturong is one of the largest and least graceful members of this family, weighing around thirteen kilogrammes, and is one of only two carnivores to be supplied with a prehensile tail. The other is the kinkajou. The kinkajou is a member of the raccoon family, the equivalent of the civets in the New World. Related creatures include the brown coati, a diurnal omnivore, and the nocturnal olingo. The small-eared fox of Amazonia and the Orinoco basin, although belonging to the dog family, is very similar to a mongoose in general appearance, with its short legs, heavy tail and long muzzle.

A second group of carnivorous mammals common in tropical forests, both in the Old and New worlds, are the mustelids. The hog badger, as its name suggests, rather resembles a wild boar. Its long muzzle serves to unearth both animal and vegetable food, and it is endemic to forests in the Oriental region. The ratel, or honey badger, is widely distributed from Asia through Arabia to Africa, whereas the Palawan stink badger, named for its ability to lift its tail and eject a nauseous secretion in self-defence, is confined to a few islands in the western Philippines. The ferret badgers of Southeast

Asia also possess these anal glands, but they are much smaller animals, weighing only about two kilogrammes. All three species have a white stripe across the muzzle, which is thought to give warning of their noxious manner of defending themselves. South American forest mustelids include the omnivorous tayra and the black and white grisons, which resemble miniature ratels and are among the fiercest of all small, rainforest carnivores.

The diversity and abundance of feline species which frequent the world's tropical forests is quite phenomenal, ranging from the diminutive flat-headed cat of the Oriental region to the undisputed lord of the jungle, the tiger. No less than nine species of smaller cat are found in Asiatic jungles, together with the larger felines – the tiger, clouded leopard and leopard. The small hunters are characterised by large eyes with narrow vertical pupils for hunting in the dark, rounded heads, short muzzles and less powerful jaws than the big cats, since most species are omnivorous. Some are very restricted in their range, such as the bay cat of Borneo and the Philippines cat, whilst others, such as the leopard cat, are distributed right across Asia. The range of the caffer cat even extends into Africa, where it is known as the African wildcat.

13

Tropical reptiles range from the massive American alligator (**5**) to the slender, wide-mouthed Central American snakes (**3,4**) of the genus *Leptophis*.

The clouded leopard is thought to be the nearest living relative of the sabre-tooth tiger. Although its canines are longer than those of most modern-day felines, they are quite insignificant compared to the twenty-centimetre-long teeth of the legendary sabre-tooth. Like its smaller cousins, the clouded leopard is an arboreal predator, active mainly at night, hunting small and medium mammals and birds. It is distinguished from the other big cats by its longer muzzle and darker coat, and is confined to Asia. The leopard is more widely distributed, inhabiting both African and Asian jungles as well as swamps, deserts and high mountain habitats. Asian leopards are frequently melanistic, the black forms being known as panthers. The forest species weigh around ninety kilogrammes and have a particular liking for monkeys, although they will also eat animals ranging from frogs and mice to young gorillas. All leopards have strong retractile claws for climbing trees, short muscular legs, a long tail for balancing and a finely spotted coat to provide them with camouflage.

The South American equivalent of the leopard is the jaguar, in which melanistic forms are also common. About the same size as leopards, jaguars are more thickset, with proportionately larger heads, and the rosettes, or ocelli, which decorate their coats have black centres. Jaguars prey on a wide range of animals, including capybara, peccaries, monkeys, deer and

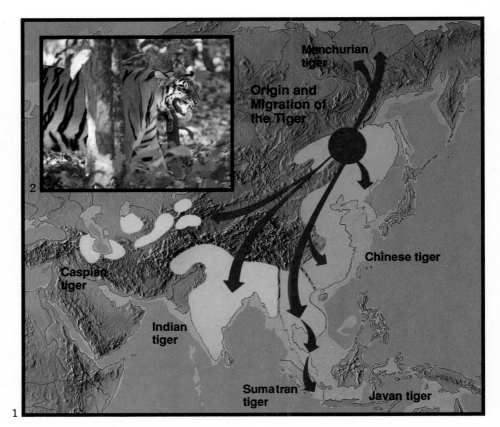

tapirs, as well as on fish and caimans. Other New World felines include the ocelot, jaguarondi and margay, all of which are considerably smaller that the jaguar. The jaguarondi, with its short legs, slender body and small, flat head, is also known as the otter cat. It is the only forest cat which is not spotted. Individuals may be black, russet or grey; a typical litter will contain kittens of each colour.

The largest living feline is the tiger, approaching four metres in length, almost a quarter of which is tail, standing around ninety centimetres high at the shoulder and weighing up to 275 kilogrammes. Most tropical forest cats (other than the jaguarondi) are either spotted or striped from head to tail, but the tiger's magnificent coat possesses stripes running from the middle of the back towards the ground. The pattern varies according to the subspecies and also with each individual, but the stripes are never symmetrical.

Like most jungle cats, tigers are nocturnal, feeding mainly on large terrestrial herbivores. But instead of leaping down onto their prey in the manner of a jaguar or leopard, the tiger hunts by running into its selected victim at full speed, bowling it over and then sinking its fangs into the neck. This has the effect of snapping the vertebrae of small animals and will eventually suffocate

The ocelot (**6**), found from the southern United States to Paraguay, is a small, short-legged, nocturnal cat which swims and climbs well, while the not dissimilar African palm civet (**8**) is in fact a member of the mongoose family.

Tigers (**2,5**) are thought to have originated in northern Siberia, working their way south some time after the last ice age via two main routes on either side of the Himalayas, gradually adapting to the warmer climates and different habitats and diverging into a number of distinct races (**1**). The speed of the tiger's spring (**3**) spells instant death for most intended victims, but if necessary the tiger can sprint over short distances to bring down its prey, in this case a monkey (**4**). The tiger's main prey are deer, antelopes, wild boar, rodents, fish and domestic animals, and even man on occasion.

7

8

9

10

a larger beast. As a rule, tigers prefer young or small mammals, but have been known to attack water buffalo and even crocodiles and turtles. The rare tiger that attacks man is usually an old individual, whose worn-down teeth and slow reflexes have reduced its prowess in hunting more wary prey.

There are eight recognised subspecies of tiger, three of which – the Caspian, Balinese and Javan races – are, sadly, extinct, and the future of the others is by no means assured. It is estimated that in 1930 there were over 100,000 tigers in the world, including

representatives of all eight species. Yet by 1970 both the Caspian and Balinese species had vanished forever; a handful of tigers were just surviving in Java; the Chinese race was thought to be represented by only a few individuals, and the Siberian tiger, the largest of all, had been reduced to some 130 beasts. The total number of tigers at this time was estimated at around 5,000. Since this time, as a result of the success of Operation Tiger and the establishment of a network of reserves, the numbers of Sumatran and Indian tigers have almost doubled, and the Siberian tigers amount to

Leopards (**7,9,10**) are the most widely distributed of all the big cats, both geographically, from Africa to China, and in terms of habitat, being equally at home in deserts, mountains, savanna grasslands and tropical forests.

between 350 and 400 individuals. However, the Javan tiger has finally succumbed to the pressures of habitat loss and hunting.

The plight of the tiger is by no means an isolated one and, for some species, recognition of their vulnerability has come too late. Many others undoubtedly become extinct without ever having been discovered. The localised distribution of many tropical forest animals and plants makes them especially vulnerable to disturbance, and the complicated interrelationships of forest species means that the loss of one may lead to the extinction of many more. A conservative estimate is that one rainforest species vanishes forever every day, although some sources

1

Brilliantly-coloured plumage, often iridescent, is a feature of tropical forest bird-life that is vividly illustrated in the South American macaws (**2**) and the bee-eaters of Eurasia. This Ceylon green bee-eater (**6**) is an immature bird that has not yet attained full adult plumage.

2

3

4

Many rainforest reptiles are gaudy, but since the vegetation is so lush a bright green snake such as the emerald tree boa (**1**) is less conspicuous in ambush than might be imagined. The bloodsucker (**3**), a lizard of the genus *Calotes* belonging to the Agamidae, or chisel-teeth lizards, can change colour readily during courtship or combat, the head often becoming bright red, a fact from which this elegant reptile derives its name. Chameleons (**4**) are also renowned for their colour-changing ability, although in their case this is prompted more by a change of background and the need for camouflage. (**5**) A python well camouflaged in water-logged leaf litter.

suggest that this figure is approaching one per hour. And it is not only wildlife that is suffering; almost a hundred Amazonian Indian tribes have disappeared since the beginning of the century, and with them has vanished the folklore that in the past has guided scientists to plants with medicinal properties.

In the last decade, more than two and a half million square kilometres of tropical forest have disappeared, and the destruction continues at a rate of around 200,000 square kilometres per year. Slash and burn agriculture, fuelwood collection by impoverished local people, commercial timber extraction by multinational companies for furniture, paper and energy production and clearance of the forest by fire to provide grazing for cattle – largely to supply the needs of the fast-food

5

industry – have undoubtedly had the biggest impact. And yet, as each hectare of forest falls to flames and chainsaw, many potentially valuable species disappear, their worth never recognised. Pharmaceuticals derived from tropical forest plants, such as the cancer-battling drug vincristine, extracted from the rosy periwinkle, save thousands of lives each year and generate billions of pounds worth of industry annually in the Western world. Coffee, bananas, rubber, essential oils, vegetable waxes and resins are all derived from the rainforest, and there are around 1,650 known tropical forest plants which have potential as vegetable crops.

Even more importantly, tropical forests are thought to contribute greatly to global climatic stability. The burning of large areas of forest releases vast quantities of carbon dioxide into the atmosphere, the knock-on effects of which – global warming, melting icecaps and rising sea levels – are only just beginning to be realised. In addition, forests hold large quantities of water which is released gradually into the atmosphere from the leaves under normal conditions. The removal of the trees is almost certainly responsible for the floods and

6

7

8

droughts which alternately plague deforested regions. Likewise, the nutrient store of a tropical forest is effectively destroyed by the removal of the vegetation, either by burning or logging, and the once lush growth of the rainforest is never repeated in crop yields or even in the secondary forest which may appear if the land is abandoned.

It is estimated that 5,000 years ago about fourteen per cent of the earth's land surface was covered with the dense green mantle of tropical forests. Today, less than half of that percentage remains. Most of it has been lost in the past 200 years and, at the current rates of destruction, many scientists predict that another fifty years will see the total demise of this, the world's richest living resource.

The tiger hunts mainly by night, resting in the forest shadows during the day, its stripes rendering it near invisible (7). Young South American tapirs (8) are also striped and spotted, in this case to avoid detection by predators, especially jaguars.

CHAPTER 11

TEMPERATE FORESTS

The temperate forests of the world vary considerably in species composition and structure but are linked by their seasonal aspect. From the towering temperate rainforests of southeast Australia to the ancient deciduous woodlands of central Europe, they support a wealth of animal and plant life which is uniquely adapted to the changes in temperature and precipitation that take place during the course of the year. The value of the fertile soils produced by these woodlands has long been recognised by man, with the result that vast tracts of forest have been replaced by agricultural land, and in many areas only fragments of this temperate splendour remain.

Left: a white admiral butterfly, (top) a song thrush's nest, with its clutch of sky-blue eggs, and (above) a badger.

Temperate forests represent the natural vegetation type of huge expanses of the northern hemisphere, bounded by the dark green band of coniferous forests to the north and separated from the tropical forest zone at the equator by extensive grasslands and deserts. They differ from tropical forests in that they grow where there is a marked cold season, when temperatures drop well below freezing point, or where a seasonal drought limits the availability of water and so curtails tree growth. In some parts of the temperate world the trees respond to these unfavourable conditions by shedding their leaves, and are thus known as deciduous species; in others, the trees are predominantly evergreen.

There are four basic types of temperate forest in the world, the most northerly of which is almost a transitional stage between the deciduous temperate woodlands and the Boreal forests. It is commonly composed of mixed stands of broad-leaved deciduous trees and coniferous species and probably represents the climax vegetation type of much of northern central Europe,

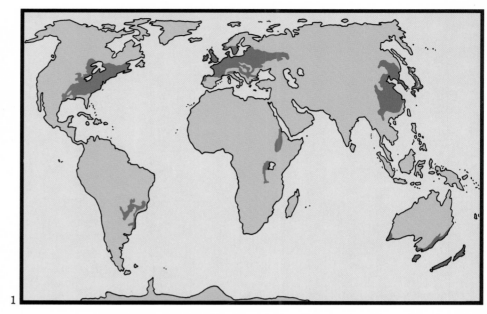

1

2

3

eastern Asia and northeast America, although only fragments remain today. The temperate forests of the Appalachian Mountains in North America, for example, comprise mixed coniferous formations of hemlock, white pine, beech, sugar maple, yellow birch, black cherry and oaks.

The second type comprises a mixture of conifers and broad-leaved evergreen species and was formerly common across many regions with a Mediterranean climate, although little remains today due to the long occupation of these lands by man. Evergreen oak forests once covered more than two-thirds of the Iberian Peninsula, although the natural Mediterranean vegetation is so fragmented today that it is almost impossible to know what its component species would

be in undisturbed conditions. On the few steep slopes where Mediterranean woodlands persist, the range of evergreen, sclerophyllous shrubs is enormous – Mediterranean buckthorn, many fragrant species of cistus, wild jasmine, strawberry tree, wild olive – growing with conifers such as junipers, stone and maritime pines and Spanish fir in a few localities. Some deciduous species, such as lentisk, turpentine tree, barberry and Mediterranean coriaria, are also present.

The third type of temperate forest consists almost entirely of deciduous trees and once covered much of Europe and northern Asia. Deciduous Mediterranean woodlands, for example, are dominated by Lusitanian, Pyrenean or white oaks, together with limes, Montpelier maple and sweet chestnut, while the Apennines of

Temperate forests are found in both hemispheres (1), between the tropics and the tundra regions. Within these shady retreats bloom delicate white helleborines (5), the rare and exotic lady's slipper orchid (7), and a whole host of fungi, such as the tiny, fragile *Mycena crocata* (6), confined to the beechwoods of England's North Downs.

4

Europe's ancient temperate woodlands support a wealth of animal life, although the shy nature of these creatures almost belies their existence. Sunny glades that are occupied by handsome purple emperor butterflies (3) and watchful roe deer (4) by day become the haunt of the rodent-catching tawny owl (2) at dusk.

5

6

7

central Italy still support extensive forests of beech, birch, yew, hop-hornbeam and flowering ash, the exact composition of which varies mainly with altitude. The deciduous forest belts of the Soviet Union, on the other hand, are largely composed of pedunculate oak, ash, Norway maple, elms, small-leaved limes, pear and apple, with beech and hornbeam to the west. They are still in a relatively natural state as, lying in a region where population densities are very low compared to those in the rest of the northern hemisphere, large areas have remained almost untouched by man. Similarly, eastern Europe still boasts some very ancient, virtually undisturbed temperate forests, such as that of Bialowieza on the Polish/Soviet border. The core of the national park on the Polish side, which amounts to only about forty-eight square kilometres, has been left to its own devices for several decades, and its towering oaks and pines and large numbers of fallen trees probably come close to recalling the ancient forests of Europe.

The fourth temperate forest formation in the world is characterised by very high levels of rainfall, producing a vegetation type remarkably similar in many ways to the tropical jungle. Although located in seasonal climates, these forests are dominated by broad-leaved evergreens and are known as temperate rainforests. They occur in both hemispheres, along coasts where incoming clouds deposit a considerable amount of rainfall during the course of the year. In many cases,

1

2

3

temperate rainforests would appear to be a continuation of the tropical semi-evergreen forests which fringe the jungle. Broad-leaved conifers such as podocarps – primitive conifers which resemble yews – are common components of the canopy, and the understorey frequently contains bamboos, tree-ferns and palms. Epiphytes and climbers are also typical, although woody lianas are for the most part replaced by herbaceous species.

In New Zealand the subtropical forests of the north are the realm of the kauri pine and podocarps, some of which grow to a height of fifty metres, but in South America the west-facing slopes of the Andes support lush forests of magnolias and laurel-like species such as myrtles, which may exceed forty metres. In Australia, temperate rainforests of another type clothe the southeastern corner of the continent, including much of Tasmania. Here the tallest trees are a species of eucalyptus known as mountain ash, some of the largest specimens of which extend up to a hundred metres in height. The understorey contains some of the most primitive woody species still in existence, such as tree-ferns, together with banksias, sassafras and myrtle

4

beeches. Along the rain-washed Gulf coast of North America, especially in Florida's Everglades, familiar forest trees such as oaks grow alongside magnolias, royal palms, mahoganies and bromeliads. Huge ferns dominate the shady undergrowth and the trees are festooned with epiphytic plants.

Other temperate rainforests occur in much cooler climates, although again usually in maritime regions. In

The Australian koala (**1**) is a eucalyptus-eating marsupial mammal that gives birth to a tiny baby that spends its early life suckling at a teat hidden deep within its mother's backwards-facing pouch. The echidna (**4**), a monotreme, is even more primitive, the mother laying a single egg directly into an abdominal pouch that forms at this time. The egg hatches after about ten days and the baby suckles for around sixty days on milk exuded directly through the skin.

Cork oaks (**6**) are reminiscent of the evergreen Mediterranean forest that once covered most of Iberia, although today they are mostly cultivated, providing cork, acorns and charcoal. The poisonous red berries of the arum lily (**5**), sometimes known as lords-and-ladies or cuckoo-pint, can be found in almost any damp, European woodland in autumn. The fox squirrel (**2**) is the largest of the tree squirrels, measuring over half a metre. It is found mainly in eastern North America, feeding on nuts, especially hickory and acorns. There are usually four red fox cubs (**3**) in a litter, born blind and helpless in March or April, but weaned at eight to ten weeks and sexually mature by December.

5

the southern hemisphere the southern beech forests of the west coast of southern Chile are among the best examples. In southern New Zealand the podocarps gradually give way to forests of southern beech, especially the mountain, red and silver beeches typical of regions experiencing colder winters. Yet another type of cool temperate rainforest, which occurs in southern Alaska, is almost entirely coniferous. Along the west-facing slopes of the Pacific Coast Range are lush forests of western hemlock, western red cedar, sitka spruce and Douglas fir. The dimensions of these conifers recall the enormous trees of the tropics, wreathed with epiphytic ferns and literally dripping with moisture for much of the year.

But the very existence of these temperate woodlands resulted in the build up over millions of years of thick, fertile soils, and early man soon discovered the value of these lands in his first tentative experiments in cultivating crops and grazing his newly domesticated livestock. In Europe the clearance of the original oak and beech forest began with Neolithic man, and accelerated in historical times due to the timber requirements of shipbuilding and house construction and the need to

6

feed the burgeoning human population. Today only fragments of this impenetrable forest, which even up to Roman times was roamed by aurochs, bison and bears, remain. As a result it is extremely unlikely that any entirely natural areas of deciduous woodland – those that have never been influenced or exploited by man – exist in western Europe today. At best, there may remain a few regions which have been continuously forested since at least the Dark Ages, and it is these which the scientists refer to as 'semi-natural ancient woodlands'.

In Britain, for example, temperate woodlands once covered about seventy per cent of the land surface, but today they account for less than ten per cent, much of which has in fact been replanted with non-native conifers to produce monocultures extending over huge tracts of upland and lowland alike. The remaining semi-natural woodlands almost all show signs of man's activities, such as old pollarded or coppiced trees and ancient banks and ditches indicative of animal enclosures when the woodland was also used for pasturing animals.

1

2

3

4

Oak trees (**3**) feature prominently in the world's temperate forests, growing mainly in the northern hemisphere, while red gum trees (**4**) are typical of the eucalypts which dominate the Australian temperate woodlands. Sweet chestnuts (**2**) have been introduced to much of Europe because of their valuable timber and for their produce.

Similarly, the temperate forests of New Zealand have been subject to clearance for agriculture or replanted with faster-growing exotic softwoods such as Monterey pines; today less than a third of their original extent remains, and much of that is in a severely degraded state.

The history of the North American temperate woodlands is essentially the same. Although clearance did not begin in earnest until the arrival of the White man, the rate at which the primeval New World forests disappeared was very rapid; it is estimated that barely one per cent of that present at the beginning of the seventeenth century is still untouched. And as the forests vanished, so too did their larger inhabitants, such as pumas, wolves and wapiti, driven into remote and inaccessible mountain regions.

These American woodlands contain a much greater diversity of temperate tree species than their European counterparts – some forty species per hectare as compared to an average of eight. This is largely due to the fact that the forests were able to retreat southwards during the Ice Ages with relative ease since the mountain chains run approximately north-south, and likewise the spread of tropical species northwards has proceeded unhampered. In Europe, however, not only did ranges such as the Pyrenees, Alps and Carpathians severely

5

6

7

winter or the beginning of a summer drought. In effect, they become dormant as a seasonal response to low ambient energy levels and a lack of available water; physiological drought occurs when the temperature drops below about 4°C as the roots cannot absorb soil water under these conditions, and thus photosynthesis cannot take place.

The Eurasian long-tailed tit (**7**), which possesses a tail longer than its body, constructs a complicated dome-shaped nest from moss and spiders' webs.

8

9

Sheets of ramsons, or bear garlic (**5**), are common in European mixed woodlands, especially on base-rich clays. Other shade-tolerant herbs include the striking toadflax *Linaria triornithophora* (**9**), found in northern and central Iberia, and the dark red helleborine (**6**), a fragrant-flowered orchid found in rocky limestone woodlands across most of Europe as far east as the Caucasus. The black stork (**1**) is a solitary, retiring forest bird that breeds mainly in trees and feeds around the margins of freshwater lakes and marshes. Australian wagtail willies (**8**) belong to the flycatcher family and are also known as fantails because of their broad tails.

impede the southerly migration of the forests, but also the Mediterranean Sea and the Sahara desert proved to be almost insurmountable obstacles to the northwards movement of tropical forest species from Africa. The richest temperate forests of all are found today in the Oriental region of China, which was relatively unaffected by the ice sheets during the last glaciations. Chekiang province, for example, boasts some 180 temperate trees, including both broad-leaved and coniferous species, whereas France possesses only a score of common forest trees.

Although evergreen trees are present in a variety of the world's temperate woodlands, the truly characteristic species are deciduous; that is, they lose all their leaves at the same time in a response to unfavourable environmental conditions, such as the approach of

Almost all deciduous trees are angiosperms, rather than gymnosperms like the conifers which dominate the Boreal forests, although a few deciduous conifers, such as larch, serve to confuse the picture. On the whole, most deciduous species are also broad-leaved and hardwoods, the former on account of their typical paper-thin leaves with a very large surface area, the better to intercept the light necessary for photosynthesis, and the latter because they produce notably harder timber than coniferous trees, or softwoods. The proportions of an average deciduous tree never equal those of the giant coniferous redwoods of North America. Their average height is around thirty-six metres, and their average circumference some five-and-a-half metres, although very ancient specimens may be considerably larger.

A deciduous tree does not usually produce flowers and fruit until it has been established for a couple of years, and even then it may not do so every year, depending on environmental conditions. Most deciduous trees produce wind-pollinated flowers, especially in the form of catkins, as in species such as hazel, birch and alder. These often develop before the tree is in full leaf in order to assist successful cross-pollination and to give the fruit more time to develop and mature. The majority of wind-pollinated species produce separate male and female flowers which are often located on different trees, as in the case of poplars and willows. A minority of deciduous trees produce hermaphrodite, insect-pollinated flowers. This is particularly noticeable in the rose family, in species such as wild cherries, rowan, whitebeam, apples and pears, as well as in understorey shrubs such as sloe, wild roses and hawthorn. Other insect-pollinated trees include horse chestnut and lime, the former with large, showy flowers and the latter with inconspicuous blooms. In general these species flower later in the year, when more adult insects have emerged to pollinate them.

are favoured by stands of beech, hornbeam and sessile oak, while loamy soils derived from clays usually support forests of pedunculate oak, elms and limes. Soils overlying calcareous rocks such as limestone and chalk frequently support beech, field maple and ash forests, whereas waterlogged substrata are dominated by willows, sallows and alder.

1

Although fungal fruiting bodies are found in many of the world's habitats, nothing can rival the splendour of the temperate forest toadstools in spring and autumn. Many species, such as the cep (**15**) and the chantarelle (**4**), are edible, whilst others, such as the fly agaric (**16**), are highly poisonous and advertise the fact with bright warning colours. Many amanitas, including *Amanita flavoconia* (**1**), the false death cap (**3**), *A. echinocephala* (**6**) and the grisette (**12**), have mycorrhizal associations with tree roots to supplement their saprophytic nutrition. Others, including the sulphur tuft (**13**) and *Pholiota adiposa* (**5**), decompose fallen trees. Evocatively-named species include the candle-snuff fungus (**9**) and the magpie fungus (**11**). Other typical woodland fungi range from such delicate species as the purple-gilled *Leptonia euchroa* (**10**), the fragile cups of *Crepidotus variabilis* (**8**) and the feathery pink form of *Clavulina cristata* (**14**) to the more robust *Boletus erythropus* (**7**), with pores rather than gills.

2

3

The seeds of deciduous trees are either wind- or animal-dispersed. Wind-dispersed seeds are light or have a large surface area, such as those of willows, with tufts of hairs which act as a parachute, or those of the various species of ash and maple, which have broad wings to spin the seeds far from the parent tree. Seeds designed to attract animals are sometimes fleshy, such as cherries and apples, with a tough, indigestible coat protecting the seeds within. But the majority of deciduous trees produce hard, dry seeds, including acorns, beech-mast and chestnuts, which are favoured as an autumn and winter food supply by many woodland animals. In the process of being collected, consumed or stored, many of these seeds will be dropped in transit, to germinate when suitable conditions return.

As a rule, deciduous trees prefer fertile soils, reasonable levels of warmth and sunshine and an ample supply of water during the growing season. But the various species have different tolerances of such environmental factors, especially soil parent material, a fact that results in the development of large areas of forest dominated by one, or maybe two, canopy species. In Europe, for example, freely-drained soils

4

5

The composition of a woodland also depends on the successional stage that it has reached. Young woodlands, for example, where more sunlight is present, often have a canopy dominated by such light-demanding species as birch and ash. This provides shelter for the seedlings of slower-growing, but ultimately more competitive, species such as oak, hornbeam and beech, which will eventually out-shade the opportunists and come to dominate the climax woodland; gaps in the canopy of mature woodlands are subject to the same successional processes. A whole range of environmental conditions will determine the final composition of the mature woodland, but birch is rarely found as a dominant component of the climax community except where the poor soils and freezing winters of the far north preclude the growth of more competitive trees.

The most distinctive features of any temperate woodland are the changes that it undergoes during the

The woodcock (2) is one of the few waders found away from freshwater or coastal habitats, both sexes being superbly camouflaged in order to incubate their eggs on the forest floor.

course of a year, changes that affect not only the trees but all forest inhabitants. At the onset of unfavourable conditions the chlorophyll in the leaves of the deciduous trees starts to break down, revealing colourful pigments which are normally masked by the green coloration during the growing season. Where the leaves contain high concentrations of carotenoids the trees turn orange or yellow in the autumn, while those in which anthocyanins predominate become purple and deep red. Finally a thin layer of corky tissue forms across the base of each stalk, separating the leaf from the branch on which it grows, and the first strong wind sees hundreds of brightly-coloured leaves torn from the trees, eventually coming to rest on the forest floor.

The fallen leaves of deciduous trees normally decompose within a couple of years, but those of some species, such as beech, are protected by thick cuticles and a high tannin content, and as a result take much longer to break down. For this reason, the floor of a beech wood is often covered in a thick, crackly layer of several years' leaf-litter, which permits few plants to grow. Normally, however, the millions of tiny soil creatures launch their attack on the fallen leaf the minute it reaches the floor, breaking the tissues down into their component molecules, which are later reabsorbed by the roots of trees, shrubs and herbs when favourable growing conditions return. In comparison with a tropical forest, however, in which almost all nutrients are tied up in the living biomass, a much greater proportion of the total nutrients present resides in the soil of a temperate woodland at any one time.

Seasonal variations in tree cover also affect the vegetation that grows on the forest floor. So well adapted is the herbaceous layer that many plants have evolved a matching rhythm, starting their spring growth and producing their flowers before the trees are in full leaf, when the amount of light filtering through the branches is considerably higher than at the height of summer. A European deciduous forest in spring is carpeted with all manner of pre-vernal herbs: great

5

6

The Distribution of the North American Raccoon and the Crab-eating Raccoon

8

North American Raccoon

Crab-eating Raccoon

7

The redwing (**4**) is one of Eurasia's most migratory thrushes, with a breeding range almost completely discrete from its more southerly wintering quarters. Shy bullfinches (**6**) also winter to the south of their Eurasian breeding sites. The arboreal, prehensile-tailed Virginia opossum (**5**) is North America's only marsupial, producing a maximum of three litters of up to a dozen young each year.

9

10

sheets of bluebells and primroses, shy violets and wood-anemones, as well as hepaticas, wood sorrel, ramsons and sanicle. Other species, such as dog's mercury, helleborines, bird's-nest orchids, Solomon's seal and martagon lilies, are more tolerant of low light levels and are able to flower when the canopy has closed completely. The majority of woodland herbs also undergo a corresponding die-back in winter, storing energy for the coming spring in underground tubers or bulbs.

For the most part the teeming life of a temperate deciduous woodland in summer is not in evidence during the winter. Invertebrates elect to spend the winter in a dormant state, usually as pupae, although

some will overwinter as eggs, larvae or even adults, and the lack of available food thus forces insectivorous birds to find winter feeding grounds elsewhere, promoting mass migrational movements. Seed-eating birds, on the other hand, are not obliged to move on at the onset of winter and, indeed, many birds expressly visit deciduous woodlands at this time of year, taking advantage of the rich food supply therein. In Europe, bramblings descend on the beechwoods to feed on the copious quantities of beechmast, hawfinches remain in their hornbeam woodland habitats all year round, whilst thrushes, especially fieldfares and redwings, fly south from their Arctic breeding grounds, taking shelter in the woods and feeding on the fruits of such trees as

Light-demanding plants that bloom before the trees come into leaf include the butterbur (**3**), whose sixty-centimetre-long leaves appear later in the year, and the green hellebore (**9**). Later-flowering species include the shade-tolerant, highly fragrant lily-of-the-valley (**2**) and large-flowered hemp-nettle (**10**), which thrives only on woodland margins.

holly and yew, which also provide a vital winter food supply for the magnificent capercaillie in its southern European haunts.

Small mammals, such as dormice, hedgehogs and woodchucks, incapable of travelling over huge distances, elect to stay in the forest. Most hibernate, entering a state in which their heartbeat is slowed and their respiratory and metabolic rates are both dramatically reduced; the heart of a woodchuck, for example, beats about eighty times per minute during the summer, but only four times in the same period during its long winter hibernation. Prolonged spells of good weather can awaken these small forest mammals, and they may emerge briefly into the sunlight to feed, while very cold conditions provoke the same response, since a further drop in body temperature would mean almost certain death. It is for this reason that small mammals often lay in a store of food during the autumn, so that they are able to become active, feed and thus raise their body temperatures to stave off the cold. Larger mammals, especially browsers such as deer, remain active all year round, nibbling at the branches of shrubs and the

1

2

3

4

5

permanent layer of mosses and lichens when all other vegetation disappears for the winter. Bears, like small mammals, also enter a state of 'suspended animation', although in their case it is not true hibernation, instead resembling a deep sleep from which they are easily awoken if disturbed.

Like their tropical counterparts, temperate forests have a well-defined vertical structure, although the trees are on the whole smaller. In Europe the canopy – the uppermost layer – rarely exceeds thirty metres, and may be only one-third of this height. It is composed of the leafy crowns of such species as oaks, beech, elms, sycamore, limes and sweet chestnut, and absorbs a good deal of the sunlight during the summer. In North America, lindens, basswoods, maples, hickories and tulip trees add to the diversity of tree species present in the canopy. The shrub layer beneath is dominated by evergreens such as holly, box, juniper and yew, together with deciduous species which can tolerate low levels of light, especially hazel, elder, hawthorn and guelder rose, as well as the saplings of the canopy species themselves.

Mature beechwoods, however, have an extremely efficient leaf arrangement which intercepts the

6

Noisy, gregarious jays (5) feed on anything from acorns to carrion, while more specialised predators include the buzzard (3), one of Eurasia's commonest diurnal raptors, and the magnificent eagle owl (9) of the Old World, capable of taking a young roe deer or adult capercaillie. The bizarre Australian tawny frogmouths (7), related to nightjars, are less agile predators, being clumsy fliers and feeding mainly on fledglings, crawling insects and terrestrial mammals.

7

Epiphytes are less profuse in temperate deciduous forest than in the tropics, although trees growing in regions of ample rainfall are often shrouded with algae, mosses and lichens and the wider branches and forks provide a foothold for small ferns such as polypodies. The beechwoods of northern Spain's Cordillera Cantábrica, for example, are festooned with streamers of a lichen misleadingly known as Spanish moss. Deciduous woodlands also feature climbers, the most common of which in Europe are various species of ivy,

Beech trees cast a very deep shade in summer, but the young leaves (2) are pale and translucent, letting more sunlight through and allowing herbs such as the large white helleborine (1) to flower.

8

9

10

11

Small mammals of temperate forests include the fat-tailed, or edible, dormouse (4), an arboreal, nocturnal rodent of central and southern Europe and southwest Asia, and the related garden, or oak, dormouse (11), especially common in Mediterranean woodlands, spending much of its life on the forest floor. Red squirrels (6) are excellent climbers, possessing rudimentary thumbs on their forefeet and long, bushy tails to aid their balance; yellow-necked mice (8) are also frequently encountered in the canopy. Sika deer (10), closely related to red deer, are native to broadleaved woodlands in eastern Asia, but have also been introduced into Europe and North America.

maximum amount of light for photosynthesis and casts a heavy shade. This lack of light and the slow rate of decomposition of beech leaves conspire to produce a very poor understorey and ground flora and, except for the shade-tolerant seedlings of the beech trees themselves, beechwoods are almost devoid of plant life. Some very specialised herbs, such as the ghost orchid, are unable to grow anywhere else, but even these are so sporadic in their habits that decades may pass without a single flowering spike putting in an appearance. A beechwood in summer, in contrast to most deciduous woodlands, consists of just one level – the canopy.

bryony, honeysuckle and clematis, weak-stemmed species which clamber up the existing vegetation by means of hooked thorns and tendrils in a bid for sunlight.

The trees, shrubs and various herbaceous plants of the temperate forest provide a wealth of food for invertebrates at all stages of their life-cycles. Some species feed on the leaves themselves, others sip nectar from the spring flowers or eat the ripe fruits, yet others burrow beneath the bark of the trees or spend their whole lives in the mulch of decomposing plant material on the forest floor, whilst a multitude of species live only in rotting wood. It has been estimated that up to one-fifth of all woodland invertebrates inhabit this dead wood habitat, and as such are only found in old, undisturbed woodlands where fallen trees are left to decompose rather than removed for timber or firewood.

The oak tree plays a particularly important role for woodland invertebrates. Research has indicated that upwards of 300 insect species may be found on a single living tree, each occupying a specific niche. Many feed directly on the various parts of the tree itself, in turn providing food for carnivorous species and parasites. Such is the natural equilibrium of the oak tree food web that the herbivores are held in check by the predators and the health of the tree is rarely at risk.

Some of the most attractive insects of temperate forests are the woodland butterflies. In Europe, speckled woods frequent the sun-dappled glades, as do woodland browns and silver-washed and pearl-bordered fritillaries, while the canopy is home to white admirals and the tops of the trees are the haunts of purple emperors and white-letter hairstreaks. The eggs of each species are laid on a specific food plant. The caterpillars of the white-letter hairstreak, for example, are commonly found on elms and limes, those of the purple emperor on sallows and the larvae of the white admiral on honeysuckle. In response to such predation the trees and shrubs produce a whole range of chemicals, such as tannins and alkaloids, in an effort to deter the voracious caterpillars.

The vast array of woodland invertebrates, be they plant-eating or predatory, provides food for creatures

Mixed grassland and oak woodland (3) provides an ideal habitat for the badger (5), a large, stocky mustelid found across much of Eurasia that lives in an underground burrow known as a sett. Two or three cubs (4) are born in the spring, remain in the sett for six to eight weeks, are weaned at twelve weeks and stay with the sow (2) until the winter.

at higher trophic levels. Those living in the leaf litter and rotting wood are eaten by the amphibians – mostly toads and salamanders – which also thrive in these moist conditions, as well as by other creatures of the forest floor, such as lizards, shrews, moles and hedgehogs. On the other hand, the abundant supply of leaf-eating insects, especially soft-bodied caterpillars, is the nutritional base on which a whole range of forest-breeding birds raise their young.

Temperate forests are very rich in birds, both in terms of diversity and abundance, although the overall composition of the woodland avifauna changes during the course of the year. The temperate evergreen forests of the Mediterranean basin perhaps support a greater range of breeding bird species than any other woodland type in Europe. The striking feature of these birds is their bright plumage, often in both male and female birds. Golden orioles, rollers, bee-eaters, woodchat shrikes, hoopoes and black-eared wheatears are all common denizens of these evergreen oak woodlands. In the temperate rainforests of southeast Australia, equally colourful male superb lyrebirds display their magnificent tail feathers in an elaborate sexual display that takes place on a mound custom-built for the purpose.

The majority of birds which breed in the temperate woodlands of the northern hemisphere are, however, much less conspicuous. The females of the different species are often remarkably similar to each other, being drab, brown-grey birds able to remain almost invisible to predators while on the nest, thus ensuring the continuation of the species. For the most part the males declare their territories vocally, since a bird's sense of smell is notoriously poor and the dense vegetation precludes the use of sight, though the flash of red on a robin's breast or the contrasting yellow and black plumage of the golden oriole may also serve to warn off rivals. Tits and warblers, thrushes and finches all have distinctive songs that are recognised by members of the same species but ignored by other birds as the varied feeding techniques of forest birds means that there is little interspecific competition.

Thus any temperate woodland in spring is filled with the liquid voices of hundreds of male birds, each singing his heart out. Later in the year, when the chicks are almost full-grown, the forest will be strangely silent. Far from celebrating the joys of spring, these songs serve to warn rival birds that a certain part of the wood is the exclusive food-gathering domain of a single male. Intrusions result in scraps between the males along the borders of the territory, with the result that the fittest bird will stay in residence. Once the territory has been established, the male bird will carry on singing regardless, this time to attract a mate. In years of plenty these territories are often quite small and almost contiguous, but when food is scarce they become much larger, and only the healthiest birds will occupy territories, nest and produce offspring, thus safeguarding the genetic future of the species as a whole.

The European robin (**8**) selects a well-concealed nest site for its five to seven eggs, feeding its fledglings on soft-bodied insects. Acorn woodpeckers of North America bore holes in standing deadwood in which to store acorns for the winter (**1**). The long-eared bat (**10**) occurs in wooded areas throughout Europe, except northern Scandinavia, emerging shortly after sunset to trap flying insects.

6

7

8

9

10

White admirals (**6**), purple emperors (**9**) and silver-washed fritillaries (**7**) are all typical Eurasian butterflies of old lowland forest, flying from June to August. Their larvae feed respectively on honeysuckle, sallows, and violets or wild raspberry, all of which are characteristic woodland plants.

Pairs of snake-eating short-toed eagles (**3**), which have scaly legs and feet to protect them from the venomous fangs of their victims, rear a single chick (**1**) that leaves the nest at about two-and-a-half months. The black-plumaged imperial eagle usually rears two young, the juveniles gradually acquiring brown feathers as they mature (**2**).

Cicadas (**5**) are large homopteran bugs which live in trees or shrubs, mainly in tropical regions. The males vibrate small membranes on the sides of the body to produce the characteristic night chorus. Greenfinches, found throughout Europe and northern Africa, line their well-concealed, carefully constructed nests with feathers before laying four to five speckled white eggs (**4**) in the spring.

The vertical layers of the forest provide a variety of nesting sites, and each species tends to favour a certain niche. Buzzards, herons and eagles nest in the uppermost branches, often completely dominating the treetop, blackcaps, garden warblers and jays favour the shrubbery beneath the canopy, while nightingales and wrens inhabit dense, often thorny scrub nearer to the ground, and woodpeckers and owls nest in holes in the tree-trunks. Some species, such as woodcock, nightjar and capercaillie, even nest on the forest floor. Cuckoos, of which there are some forty species in Eurasia alone, have learned to avoid the arduous process of nest-building and rearing young altogether, depositing their eggs in the nests of other species instead.

5

Honey buzzards, named for their habit of raiding bees' and wasps' nests for the larvae, introduce their young (**7**) to this delicacy at an early stage. True buzzards (**6**) feed their two to three chicks on small mammals, and later train them to hunt before the youngsters leave to establish their own territories.

6

7

8

9

The cuckoo's host species is almost always a small passerine such as the whitethroat, although the great spotted cuckoo does parasitise the nests of crows, and it is remarkable that the cuckoo egg, although much larger, closely resembles those of the host species in both colour and markings, and the parent birds do not seem to notice the addition. Once the eggs start to hatch, the larger, pinker and more gaping bill of the baby cuckoo stimulates the adult birds to feed it to the exclusion of their own young, which gradually starve to death. Not content with such a slow elimination of its rivals, however, some young cuckoos (depending on the species) undertake to eject both eggs and chicks from the nest, balancing them on their back and levering them over the edge. The host birds continue to feed the rapidly growing cuckoo, though it is far larger than they themselves, until it is fully fledged.

The utilisation of a wide range of nest sites helps to alleviate competition between the various bird species, thus making full use of the various resources of the temperate forest. The same rationale applies to their preferred feeding zones and techniques. Treecreepers and nuthatches work their respective ways up and down the boles of larger trees, extracting invertebrates from the cracks in the bark with long, thin beaks, acrobatic titmice and chickadees hang precariously

from the branches of alders and birches, picking insect larvae from the leaves, while flycatchers and shrikes prefer to trap adult insects on the wing, launching themselves time and again from a favoured perch at the edge of a forest glade. Both blackbird and woodcock forage in the leaf-litter, although the longer bill of the latter enables it to probe deeper in search of food, especially earthworms, again avoiding competition.

Some of the best-adapted woodland birds are the woodpeckers, of which there are some 230 species, mostly confined to the Old World, although some, such as the rare ivory-billed woodpecker, the acorn woodpecker and the yellow-bellied sap-sucker, are

European cuckoos lay each of their eggs in a different nest, although a female will always select host nests of the same species, usually those of passerines such as the hedge sparrow, for her young (**8,9**).

found in the Americas. The largest of the Eurasian species is the magnificent black woodpecker, measuring forty-five centimetres from the tip of its bill to the end of its tail. As in all woodpeckers, the drumming sound it makes by rapidly hammering its bill against the trunk is the equivalent of the territorial songs of other forest birds; each species drums at a slightly different frequency, which serves to distinguish between the calls. However, some woodpeckers, such as the wryneck and the green woodpecker, or yaffle, adopt the normal vocal method of defending their territories and attracting a mate.

Woodpeckers possess a number of physical adaptations to assist them in their arboreal life. In order to cling to vertical trunks whilst feeding and drumming, the legs are short, the feet are armed with strong claws and the tail feathers are stiff, serving as a prop to support the weight of the bird. When feeding, it is thought that woodpeckers listen for the tell-tale sounds beneath the bark which indicate that invertebrates are present. Like most species, the great spotted woodpecker favours bark-boring beetles, which it

1

2

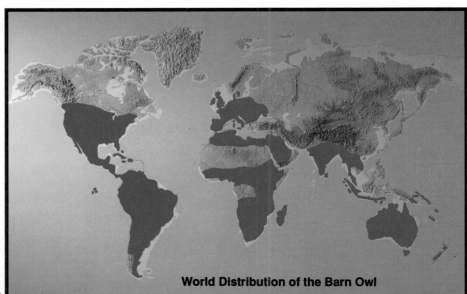

World Distribution of the Barn Owl

3

4

Owls possess a number of physical adaptations to assist them in their chosen niche. Their eyes are large and have a high concentration of light-gathering cells in the retina, and their external ear openings are also large and are asymmetrically positioned, the better to locate their victims.

5

Even the seventy-centimetre-high eagle owl (**1**) may sometimes be required to deter enemies, half-raising its wings and puffing up its feathers so as to appear larger and more threatening. Barn owls (**3**,**4**) are found on all continents except Antarctica, although in areas heavily populated by man their numbers are decreasing because of loss of habitat and the use of agricultural pesticides. Although little owls (**5**,**6**) are nocturnal hunters, they are abroad during daylight hours, whereas the tiny, secretive Eurasian pygmy owl (**7**) is rarely seen.

obtains by removing sections of bark from dead trees with its strong beak, thus exposing the holes of wood-boring insects beneath. The unfortunate invertebrates are then extracted by means of a long, barbed tongue, which may be as long as the bird's body in some species and is retracted into a chamber which coils around the back of its head.

Other woodpeckers prefer a different diet. Sap-suckers, for example, bore holes in living trees from which the sap flows freely, attracting insects to feed on its sugars. These are then eaten by the bird. The acorn woodpecker drills neat holes in standing dead wood in which it stores acorns for the winter. The most primitive member of the family is the wryneck, a reclusive bird that is seldom spotted owing to its dull plumage and secretive nature. It lacks the stiff tail of the other woodpeckers, and thus perches on horizontal rather than vertical branches, descending to the ground to feed. Because ants make up the bulk of its food the wryneck is the only migratory member of the woodpecker group.

Temperate forests also boast their fair share of predatory birds, although, owing to the difficulties of hunting down their prey between the trees, they are less numerous than in more open habitats. Generally, forest raptors are smaller than their counterparts on the plains, with short, broad wings and a long, mobile tail to enable them to twist though the trees in pursuit of their victims. Goshawks and sparrowhawks are the ultimate in forest raptor design, being able to accelerate to maximum velocity within the space of a few wingbeats. These sharp-eyed hunters spend the day perched in a tree surveying all that goes on in the surrounding forest. When dusk falls and the woodland birds start to settle down for the night, little escapes the attention of the forest hawks, who launch lightning raids on their unwary victims. The goshawk, which is by far the larger, favours magpies, crows and jays, but will also take rabbits and rodents, whilst sparrowhawks, as the name suggests, prey mainly on smaller birds.

Apart from such crepuscular hunters, temperate forests provide a suitable hunting and breeding ground for a large number of truly nocturnal predators: the owls. Owls feed mainly on small, nocturnal mammals such as dormice, shrews, field mice and rabbits, but the larger species, such as the eagle owl, may even take roe deer and capercaillie. Some species, such as the hawk owl of the coniferous forests, the snowy owl of the tundra and the forest-dwelling little owl, will venture out occasionally by day, but owls are generally at their most active as dawn approaches. Although their eyes are enormous and positioned at the front of the head, owls do not hunt by sight. Instead it is thought that they find their way through the forest at night by means of an

The female goshawk (**2**) lays three to four eggs which she incubates for up to seven weeks, leaving the nest only to accept food left nearby by her mate, and it is another seven weeks before the young venture from the nest. The North American yellow-bellied sap-sucker (**10**) bores small holes in living tree trunks to feed on the sap which is exuded, also taking any insects attracted to the sticky fluid.

6

7

8

9

10

Red foxes (**8,9**) are opportunistic woodland creatures, feeding on anything from beetles and worms to birds, mice and rabbits. They are solitary creatures, except in the breeding season, usually spending the day at earth and hunting at night.

intimate knowledge of their territories; because of this each bird has a well-defined hunting ground. Compared with other birds, owls have well-developed external ears, and their keen hearing enables them to pinpoint the least sound that would indicate the presence of a potential victim. In addition, they are able to turn their heads almost full circle, which lessens the need to move the body and possibly alert their prey, and also gives them an even more accurate fix on its position.

Once a small mammal has been located, the owl will launch itself in swift and lethal pursuit. Its primary wing feathers are fringed with fine down that enables it to swoop down on its prey in almost complete silence, and its powerful toes and long, curved talons spell instant death for the unfortunate mammal, which is normally seized by the head. During the day, owls hide themselves away in the dense foliage of the trees, camouflaged by their drab plumage. Some species, such as the North American great horned and screech owls, and the Scops and long-eared owls in Europe, have developed tufts of feathers resembling ears which serve to break up the outline of the head and render them less visible to a sharp-eyed goshawk.

The crepuscular, black-lipped roe deer (**1**) occurs in open woodlands and low mountain areas across Europe from the Mediterranean region to southern Scandinavia and into Asia. The mature bucks stand a maximum of seventy-five centimetres high at the shoulder and the does are even smaller. The tiny fawns (**2**), born in May or June, are usually twins, having spotted coats for camouflage throughout the first year.

Other Australian forest herbivores include the naked-nosed wombat, very similar in appearance to its cousin which frequents the arid plains. Since it is a subterranean creature, the wombat has a rear-opening pouch which will not fill with earth and so suffocate the young inside as the creature is burrowing. Yet another woodland marsupial is the banded anteater, or numbat. Unusually among marsupials the numbat is active by day, breaking up rotting trunks and extracting the termites which provide its main source of nourishment with its long, sticky tongue. Owing to habitat destruction and predation by introduced foxes and dingos, numbat

White-tail deer (**3**), standing up to a metre high at the shoulder, are the most familiar and abundant of North American deer, named for the white underside of the tail, which is 'flagged' as an alarm signal.

3

The mammals of the world's temperate woodlands can also be divided into hunters and hunted, although most of the larger herbivorous forest creatures have few natural enemies today since their main predators – wolves, bears, thylacines, wolverines, wildcats and tigers – have been systematically persecuted by man over the last few centuries to eliminate the threat they presented to his domestic animals.

Possibly the most bizarre assemblage of woodland mammals is found in the Australian temperate woodlands. The stocky, tailless koala, with its sleepy eyes and large black nose, superficially resembles a bear but is in fact a marsupial. Koalas are slow-moving creatures, rarely descending to the ground except to move to a new tree. They spend most of the day dozing in the crook of a branch and most of the night eating. Of the hundreds of species of eucalyptus in the forests of eastern Australia, koalas feed on the leaves of only about twenty, the first choice being manna gum. As with other arboreal creatures such as primates, the first digit of each hindfoot is opposed to the other four, but the forefeet have in effect two of these 'thumbs', giving the koala a secure grip as it moves around the tree.

4

Red deer (**4**) are found throughout Eurasia, from the Arctic Circle to North Africa, generally in dense deciduous forest, although they are also found in open moorland areas where their natural habitat has been destroyed.

numbers have declined greatly in recent years, and it is now found only in the Wandoo forests of southwest Australia.

The populations of carnivorous marsupials are even more reduced – in the case of the thylacine, there is no certainty that the creature is still in existence. The thylacine, also known as the marsupial tiger (on account of its striped haunches) or Tasmanian wolf, became extinct in mainland Australia several hundred years ago, and disappeared from New Guinea even earlier, partly due to competition from the dingo, but largely owing to persecution by man. It is now confined to the island of Tasmania, which has thus far remained free of dingos, but the last confirmed sighting was in 1961. Since that time several people claim to have seen thylacines on remote parts of the island, but it is more than likely that this, the largest flesh-eating marsupial, has vanished forever.

Although the thylacine is wolflike in overall appearance, it is thought to occupy a niche rather similar to that of the African hyena, since it is unable to hunt down its prey at speed and instead adopts a

1

2

3

4

5

6

more opportunistic life-style: that of a scavenger. Its mouth is quite extraordinary in that its jaws open to an angle of almost 180 degrees, although the advantage of such an enormous gape is not known. The thick tail is stiff and immobile like that of a kangaroo – it cannot be wagged or tucked between its legs like that of a dog – and some people claim that the thylacine uses it to balance upright on its hind legs.

Smaller than the thylacine and widespread over the continent are the so-called native-cats which are in fact carnivorous marsupials. Among those found in the temperate forests are the tiger cat, which is golden brown with irregular white blotches, and the eastern native cat, which is black with white spots on back and flanks, both of which are arboreal creatures renowned for their fierce natures. Intermediate between the thylacine and the native cats is the Tasmanian devil, also confined to that island today, although it was once common right across the continent.

On the other side of the world, the mammals of the Oriental temperate forests of Manchuria, Japan, Taiwan and Korea are no less interesting than those of Australia. Siberian tigers, the largest in the world, and leopards represent the top level in the food chain, feeding on wild boar, roe and sika deer and elk. Again, these large predators have suffered greatly at the hands of man, both through habitat destruction and direct persecution. Himalayan black bears are also denizens of these rich forests, as are raccoon-like dogs and small mammals such as hamsters and ground squirrels. The Oriental

Red squirrels (**10,11**), are found throughout coniferous and deciduous forests from Iceland to Japan and from the Arctic Circle to the Mediterranean region. The hoopoe's splendid crest (**12**) is alternately expanded and contracted when the bird is agitated.

Forests of Manchuria are also home to the mandarin duck, long appreciated in European gardens for its exotic plumage and now established in the wild in several parts of the Western world.

The North American forests have their own distinct mammalian fauna, including the American black and grizzly bears, skunks and the grey squirrel that has taken so well to Britain's temperate woodlands. Chipmunks, closely related to squirrels, are also common, spending some six months of the year in an underground chamber feeding on nuts and seeds gathered during the autumn while the bitter winter winds whistle overhead. The Virginian opossum, one of a small family of exclusively New World marsupials, is also a denizen of the North American forests, as is the raccoon, with its bushy, ringed tail and dark mask across the eyes.

7

8

9

10

11

12

Tawny frogmouths (3) resemble dry branches in their torpid daytime state. The brown honeyeater (1) has a long bill to extract nectar from even the deepest flowers. The laughing kookaburra (6), the world's largest kingfisher and probably Australia's best-known bird, will often use a smooth-barked river red gum (8) as a vantage point.

In Europe, the most prominent forest herbivores are roe and red deer and wild boar. The range of the wild boar extends right across Eurasia, from Iberia to Japan, although it is limited in the north by the severe winters and thus does not occur in most of Scandinavia and is confined to the southern reaches of the Soviet Union. Although predominantly vegetarian, rooting among the forest litter for fungi and plant tubers, or foraging for acorns and beechmast in autumn, boars have been known to take rabbits, reptiles and even fish. Both sexes are armed with four sharp tusks formed from the canine teeth, although these are better-developed in the male, and will defend themselves and their young fearlessly against wolves, their main predators, and even against man. On the whole, however, they are placid beasts and prefer to avoid conflict wherever possible.

The red deer is widespread across the northern hemisphere, although its North American counterpart, the wapiti, has a much more restricted distribution. The mature male sports a magnificent pair of antlers which come into use during the autumn rut, when the stags test their strength against each other in order to secure females and territories. These antlers are shed

All of Australia's native mammals are either egg-laying monotremes, such as the termite-eating echidna (7), or pouched marsupials. The latter include the arboreal koala (5), which sleeps all day and eats eucalyptus leaves all night, the canopy-dwelling greater glider (9), which is equipped with an extensive membrane to help it 'fly' from tree to tree, the wombat (4), whose pouch is oriented towards the rear so that it does not fill with earth during burrow excavation, and the tiger-cat, or spotted-tailed quoll (2), which is one of the largest surviving marsupial carnivores.

An attractive member of the lily family, Solomon's seal (**2**) is found in dry woodlands, especially beech woods, throughout Europe. Roe deer stags (**1**) have deciduous antlers, with up to three points and no brow tines, which are shed in late autumn and regrown in the spring.

Wolves (**4**) are found throughout the northern hemisphere, in Europe, Asia and North America, although their numbers are much reduced in Europe today as a result of human persecution. The North American grey fox (**3**) constructs a family den in a small cave or hollow tree where it hides out during the day, hunting small mammals, birds and invertebrates at night. The red fox (**7**) was introduced to North America by the European colonists to provide better sport for the fox-hunting fraternity.

every spring, and each year they become larger and possess more 'points', until in the seventh or eight year they reach their maximum development. The smaller roe deer does not extend into North America, but is ubiquitous in the forest belt which extends across the middle of Eurasia. Unlike those of the red deer, the antlers of the roe bucks are simple affairs, only about twenty centimetres long and with only a few short branches, but they serve the same purpose. Both deer species have large ears and exceptional hearing, and the females give birth to dappled fawns which are perfectly camouflaged among the shrubs of the forest.

The ground-dwelling hunters include brown bears, badgers and the red fox, all of which are omnivorous

5

6

7

8

species, while the more carnivorous predators, such as ferocious wolves, wildcats and lynxes, polecats, mink and wolverines, are equally at home on the forest floor or in the trees and are thus able to exploit a greater range of feeding territories. Small arboreal creatures – red squirrels and hazel and oak dormice – provide food for such acrobatic predators as pine and beech martens, although these opportunistic mustelids will eat almost anything, including fruit and honey. Indeed, in the face of decreasing natural habitat in Germany, Austria and Switzerland, pine martens, also known as stone martens, have taken to sleeping in the engine compartments of parked cars. Not content to use this cosy roost for shelter only, the young, inquisitive martens try their teeth on anything and everything, with the result that literally thousands of motorists every year find their cars vandalised, with wires and hoses in shreds.

Of all the inhabitants of the world's temperate forests, very few have been able to adapt to the increased human pressure on the environment. While foxes and tawny owls are now frequent denizens of the towns and cities of Europe and skunks and raccoons are not uncommon in North American urban areas, for most woodland creatures the disappearance of their natural habitat spells disaster. Even those not actively persecuted by man or hunted for sport are suffering as a result of the continued conversion of forest to

9

croplands, housing estates and industrial areas, while the replacement of the rich and varied temperate woodlands with sterile plantations of conifers to supply the softwood trade is just as damaging. There was a time when the ancient temperate forests of the world supplied man with shelter, food for his animals, fuel for his fires and a place of refuge for the spirit, where he could relax and be at one with nature. Of all these things perhaps the last is the most valuable, although its worth cannot be calculated in economic terms, and we are in danger of losing it forever.

Red deer stags (6) stand up to 1.4 metres at the shoulder, the does (9) being smaller and less heavy-set. The red deer stag's antlers bear up to twelve points, while those of the sika deer (8) are much less impressive, with a maximum of four points.

Striped-coated wild boar piglets (5) are born in spring and are able to follow their mother around after a few days, though they do not reaching sexual maturity for two years.

Above left: an amphibious mangrove crab in Indonesia and
(above right) an African lion – the king of beasts.

PICTURE INDEX